I0104676

THE BEEKEEPERS ANNUAL
IS PUBLISHED BY
NORTHERN BEE BOOKS
MYTHOLMROYD, WEST YORKSHIRE
& PRINTED BY
LIGHTNING SOURCE, UK
ISBN 978-1-908904-20-1

MMXII

EDITOR, JOHN PHIPPS
NEOCHORI, 24024 AGIOS NIKOLAOS,
MESSINIAS, GREECE
EMAIL manifest@runbox.com

SET IN HELVETICA LT BY D&P Design and Print

Front Cover
Mouse Guards *by John Phipps*

CONTENTS

1913 THORNE 2013

Supporting livelihoods and making hives for 100 years

Edgar Henry Thorne

1953

1960

1982

BEEHIVE BUSINESS PARK, RAND,
Nr. WRAGBY, LINCOLNSHIRE, LN8 5NJ
Tel. 01673 858555 email. sales@thorne.co.uk
www.thorne.co.uk

2013

FOREWORD

John Phipps

October 2012

This year has been an important and exciting one for Britain with the celebration of the Diamond Jubilee of Her Majesty, Queen Elizabeth II. Although we can only boast thirty years of publication with this edition of the The Beekeepers Annual, we are proud to have provided beekeepers in the UK with an important resource for half of Her Majesty's reign. The first Edition carried a Foreword by George Knights FRES in which he said "That this publication should take place now coincides with a new and vigorous enthusiasm and, indeed, increasing involvement which is occurring at this present time in Beekeeping". Despite all the problems which have beset the craft since George penned those lines, he could easily have been writing about the upsurge in beekeeping today. Also in the first edition, Ron Brown wrote about his experiences with the Killer bees of Brazil; Eric Milner described beekeeping in Hong Kong which was still a British Colony; Michael Coward, the then General Secretary of the BBKA, outlined EEC proposals for the bottling and labelling of honey; Robert Pickard reported on his investigations of the honeybee brain; Bill Bielby expressed his concern about the lack of beekeeping records; and Karl Showler gave us a thorough account of Charles Nash Abbot's journey to Ireland. Details of the construction and use of the Horsley Board as a valuable aid during swarming time was provided for us by Eric Hughe of Sheffield; Ken Ibbtson, of BIBBA, gave advice on producing better bees; and Michael Archer made a plea for readers to supply him with hornet samples. Adverts of that time are interesting to look back on - Steele & Brodie and Budget Beekeeping were still trading; Brinsea Products were selling a bowl-shaped, horizontal, table-top honey extractor, much like the prize-winning one which Swienty manufactured years later; cell punches for queen rearing were being sold by BIBBA; and Leaf Products were advertising foundation moulds in six sizes. The comprehensive beekeeping notes for each month were the work of Bernhard Mobus, and the delightful rural sketches which accompanied them were drawn by Janet Kerr.

The Scottish Beekeepers' Association has also had reason to celebrate, for this year marked one hundred years since the Association was formed and a special weekend convention was held in Sterling to comemmorate its founding. To give the Annual a Scottish flavour, I am pleased to include Morna Stoakley's article on Willie Smith, the inventor of the hive named after him which I enjoyed using for several years.

Whilst editing a new edition of Colin Weightman's 'Border Bees', I came across an old manuscript of Brother Adam's lecture to beekeepers in Newcastle in 1934, advice which has stood the test of time, and which I am pleased to have had permission to reproduce here. Also, from out of the past, I am including articles on Ley Lines and Alfred Watkins, the famous Herefordshire beekeeper. Whilst I myself am interested in Ley Lines as old roads, I have no interest in their supposed mystical or spiritual properties. Recently, the use of divining rods has been suggested to find the best places for hives to be set up and for the bees to flourish; having used rods in this way, out of curiosity, I can confirm that some bees have set up home in bait hives where the rods have shown considerable activity, but I would need to carry out many, many experiments to be convinced that my experiences were more than just coincidences.

Finally, another anniversary. It is fifty years since the publication of Rachel Carson's groundbreaking work, 'Silent Spring'. With all the problems facing beekeeping today as a result of what is happening in the environment, it seems that very important lessons haven't been heeded. Despite this, beekeeper numbers are increasing, and with the determined and optimistic attitude to the craft which now prevails, I am hopeful for its future.

John Phipps
September 2012

The current legal regime

Is there any GM pollen in the honey?

Yes → │ No ▼

Consignment accepted*

Is the GM crop authorised?

Yes → │ No ▼

Consignment rejected*

How much pollen is from GM crops?

▼ <0.9%   0.9%+ ▼

No GM labelling   Label as GM

*incurs high cost for test

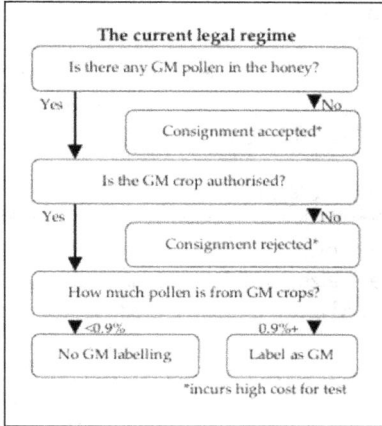

EU GM POLICY MAY DAMAGE HONEY TRADE FROM DEVELOPING COUNTRIES

Elizabeth McLeod,
Bees for Development, 1 Agincourt Street, Monmouth, NP25 3DZ, UK

The export price of Argentina's honey has fallen by 9% over the last year [1]. Many factors play a role but APIMONDIA suspects that a September 2011 ruling [2] from the European Court of Justice (ECJ) has reduced the value of Argentinian beekeepers' produce.

Top legislation from the European Council (the EU government, made up of the Ministers elected in each Member State) considered pollen a normal constituent of honey, defining honey as sugars and other substances forming honey. However, when asked to interpret European law in a politically charged dispute between an organic honey producer and a local government officer trialing genetically modified (GM) crops in Germany, the ECJ changed European policy. Reclassifying pollen as an *ingredient* in, rather than a normal *constituent* of honey, it brought pollen into the coverage of the GM regulations on food and feed. The ruling cannot be appealed because the ECJ is the highest court for questions of European law. It did not decide on the facts in the case from Germany, but its decision on the correct legal interpretation now applies across the whole European Union.

The judgment means that honey is now rejected by the EU market if it contains any trace of pollen from GM crops that are not already authorised for human consumption. Canadian honey has been removed from supermarket shelves because it is known to contain GM canola pollen. Honey imports and packers' honey stores are being tested, and contracts are being cancelled [3].

The flow chart shows how GM food and feed regulations apply to honey. It is not enough for beekeepers to ensure that their bees have no unauthorised GM forage sources. Pollen is also carried on the wind. In fact many GM crops are wind pollinated – for example maize, potato and sugar beet. If one grain of pollen from an unauthorised GM plant gets into their honey, the beekeeper cannot sell in the EU.

Authorisation of a GM crop as fit for human consumption is outside the capacity of a beekeeper who is affected by its production – it requires experimental data and an extensive risk assessment to demonstrate that the crop is as safe and healthy as a conventional product [4]. Large biotechnology firms are most successful in obtaining authorisation [5].

In any case, authorisation may apply only to use in certain food products and no authorisation of GM has been made for use in honey. For example, following the ECJ ruling, food authorities in Germany detected pollen from GM GT73 oil seed rape in South American fair trade honey. Because GT73 was authorised only for use in refined oils and food additives, they ordered the destruction of the honey and any marketing of it illegal [6].

Since over 48% of GM production was in developing countries by 2010, this ruling creates a massive barrier to their honey trade. In 2010 there were over 22 million hectares of GM crops in Argentina which is the biggest exporter of honey to the EU [7], and over nine million in India [8]. Increasingly GM is being used for pharmaceuticals and these producers have no incentive to attempt authorisation as a foodstuff [9].

Experts do not agree on whether a 10 km safety margin between apiaries and GM cultivation will satisfy EU food safety requirements [10] and approaches to enforcement across the EU are different [11] – this is creating great uncertainty in honey markets. The risk of having consignments rejected is too high for some Argentinian beekeepers who are now choosing to sell to the USA where they receive lower prices for their honey. This trend is likely to continue and to affect other honey exporters to the EU who have extensive GM cultivation, such as China and Mexico. EU producers are affected in the same way. Ultimately the use of honey in EU food products will decline and its price to consumers will rise.

While over 85% of honey imported to the EU comes from countries with GM cultivation [12] the ruling does create huge opportunities for producers of the remaining 15% (and for countries which do not currently export to the EU) to scale up production and benefit from this vacuum. If costly testing of samples was not required for such countries, this would greatly improve the incentives for them to enter the EU market and satisfy the demands of the powerful segment of EU consumers who reject GM foods.

Reprinted from "Bees for Development" Autumn 2012, by kind permission of the publishers.

References

1. NESTOR RODRIGUEZ,R. (2012) Argentina Honey Exports – June 2012. *ApiNews.* http://www.apinews.com/en/news/item/19214-argentina-honey-exports-june-2012

2. ECJ JUDGMENT (2011) http://curia.europa.eu/juris/liste. jsf?language=en&num=C-442/09 *ECJ press release on the judgment* http://curia.europa.eu/jcms/upload/docs/application/pdf/2011-09/cp110079en.pdf

3. DURKACZ J. (2012) Markets Convener for the Scottish Beekeepers' Association, *Report from the International Workshop on the ECJ Ruling on GM Pollen in Honey.*

4. http://ec.europa.eu/food/plant/gmo/authorisation/cultivation_commercialisation_en.htm

5. http://ec.europa.eu/food/dyna/gm_register/index_en.cfm

6. BECK,S. GEPA, Consequences of the ECJ judgement on the fair trade honey import to Germany (slide 20). http://ocs.jki.bund.de/index.php/GMOhoney/GMOhoney/paper/viewFile/70/36

7. WESSELER,J. Economic consequences for the worldwide trade. *International Workshop on the consequences of the ECJ judgement on GM pollen in honey.*

8. PHIPPS,J. (ed.) (2011) *The Beekeepers Quarterly* 106: 7.

9. DURKACZ J. (2012) Markets Convener for the Scottish Beekeepers' Association. Report from the International Workshop on the ECJ Ruling on GM Pollen in Honey: page 4.

10. HOFMANN,F., Working group report on Practical measures (i) for coexistence of beekeeping with commercial cultivation of GM plants and (ii) for experimental field releases, JKI Open Conference System. *International Workshop on the consequences of the ECJ judgement on GM pollen in honey.*

11. DURKACZ,J. (2012) Markets Convener for the Scottish Beekeepers' Association. *Report from the International Workshop on the ECJ Ruling on GM Pollen in Honey:* page 3.

12. Wesseler, *ibid* reference 10.

SOME POINTS ON HEATHER HONEY PRODUCTION
REV. BRO. ADAM
OF BUCKFASTLEIGH

Being extracts from the meeting of the Newcastle and District Beekeepers' Association, held in Newcastle on December 8th, 1934.

●

Being extracts from the meeting of the Newcastle and District Beekeepers' Association, held in Newcastle on December 8th, 1934.

The production of heather honey is of very ancient origin. It was known to the ancient Greeks and is referred to in old legal documents in Britain, the term "herds of bees" referring to the taking of bees to the moors.

To build up a stock for a honey-flow occurring in June or July is a simple task. It is in the nature of bees to develop in strength after winter is past. The heather-harvest, however, coincides with a stage in the life-cycle of a colony when preparations are in progress for maintaining its existence by rest, not activity. The problem, therefore, is to delay, offset, if we can, premature preparations for winter. Various means calculated to attain this are used. For instance: artificial stimulation between clover and heather harvest; doubling of colonies; strengthening by the addition of driven bees; alternatively, by contraction of the brood-nest, are some of the means resorted to, to counter-balance the decline in colony strength.

It is not my intention to enlarge on expedients of this kind for preparing stocks for the moor. We endeavour to solve the difficulty under consideration of strains which maintain colony strength until late in the season, and in unrestricted, full-development of stocks.

Strain

As just indicated, the variety or strain of bees kept is one of the foremost factors conducive to success in the production of honey on the moor. It must be realised that bees which do excellently early in the season will not perforce do well also on the moor. Conditions in general are vastly different during a heather honey-flow than those prevailing earlier in the season. Taking everything into consideration, Italians give the most satisfactory returns on the clover; but on the heather, especially the progeny of imported queens, fail notoriously. Carniolans too, although alpine bees, do not in our estimation, possess the characteristics demanded for heather honey production. We have given this variety a prolonged trial, on an extensive scale, but were eventually compelled to discard them.

We find black bees, imported from the south of France crossed with Italians, most excellent for the moor. Dutch bees seem pre-eminently adapted for the production of heather honey. Here we call them "Dutch bees" but on the Continent they are known as "heath bees." The heath bee is a variety of the common brown bee. It has been evolved, in the course of hundreds of years, by the influence of moorland environments and the most intensive system of heather honey production known. Heath bees are hardy, long-lived, good comb builders, and forage almost in any weather. They are the bees exclusively developed by heather men for heather honey production. Unfortunately it is very difficult to obtain queens of the true heath bees.

Sections

There seems almost an unlimited demand for heather honey in the comb. Moreover, well-filled heather sections command the highest market price of any honey produced, either in this or any other country. Perhaps hardly any other article of food looks as attractive as heather honey in the comb. The pure white cappings, so peculiar to heather honey, impart to sections a most exquisite delicacy of appearance.

Yet, in spite of the high price obtained, it is most difficult to produce them economically in average seasons. The problem consists not so much in getting sections filled, as in procuring complete sealing of the combs. Heather secretes nectar in abnormally low temperatures, and when such conditions prevail, bees cannot elaborate wax for sealing combs. Indeed, in seasons of this kind, the bulk of the crop obtained is generally not sealed,

Instances are on record of takes of one hundred and eighty completed heather sections from one colony. However, such returns must be attributed to exceptional seasons, or to abnormal circumstances. Drifting probably was the cause of such takes.

In working for heather honey it is imperative to have every section filled with drawn comb before being taken to the moor.

The reluctance manifested by bees, even under ordinary conditions, to build comb late in the season is notorious. This trait is greatly intensified on the moor owing, as already pointed out, to the low day and night temperatures which then so frequently prevail. In normal seasons it is next to impossible to get bees to draw out foundation in shallow frames on the moor, and consequently they will still much less build comb in sections.

The section-racks we use are designed to take one shallow extracting-comb on either side, with the frames holding the sections between. This arrangement to a great measure minimizes the number of unfinished sections; and if, when the central sections near completion, they are transposed with those next to the extracting combs, all can be got evenly filled and sealed.

When working for heather honey in shallow combs, even in the poorest seasons, some surplus is usually obtained, but to secure a remunerative crop of comb honey, a season above the average is demanded. With sections, there is the further disadvantage that all unsaleable ones have to be destroyed.

Building Up Stocks for the Moor

The crop from the heather, in nine out of ten seasons, is gathered between mid-August and September 10th. Therefore, when building up stocks destined for the moor, it is well to remember that, from the moment an egg is deposited in a worker cell until a bee emerges and assumes duties outside the hive, normally five weeks elapse. During the active season the life of a worker-bee usually does not exceed four weeks. Hence, from the eggs a queen lays between June 20th and July 10th will emerge the bees destined to gather the heather crop. Consequently, if from June 20th to July 10th breeding is interrupted, or retarded by manipulations, swarming, re-queening, accidental loss of queen or any other cause, such colonies are bound to fail on the moor. If a serious interruption of brood-rearing took place during the vital building-up period, a stock may appear strong when taken to the moor, but will subsequently dwindle rapidly and prove useless for producing a surplus.

In this connection there is another aspect on which too much stress cannot be laid, namely, the detrimental after-effects to a colony of an insufficient field force to enable the incoming nectar to be stored in the supers. A weak colony usually clogs the brood-nest with honey, and thereby endangers its survival through winter by reason of an insufficiency of young bees.

I will cite a concrete instance, in illustration of what I have just stated.

Last year two otherwise successful beekeepers placed their hives close to some of our apiaries. As a matter of fact, their bees were in one of the best districts on Dartmoor. I was rather interested, for the sake of comparison, to hear, subsequently, how they fared. Both thought they had done well, for they obtained a surplus of about 27 lb. per hive.   However, to my amazement, they reported that after August 12th their bees made no further gain; whereas, in reality, the heaviest flow occurred from the 25th to the 28th of August. Indeed, the flow was so heavy that our own stocks must have averaged about 14 lb. each day. The explanation is simple if you will recall what was said regarding the critical three weeks, from June 20th to July 10th, in the development of colonies destined for the production of heather honey. One of the two beekeepers mentioned, moreover, lost his bees the following winter.

Although the brood-chamber we have is the largest in use, yet we have to let our colonies run short of super room to compel them to store enough honey in the brood-nest for winter. It is a thing well worth remembering that the stronger a colony, the less the danger of an undue amount of honey finding its way into the brood-chamber—provided, of course, adequate super room is given in time.

Annual Re-queening

There is an idea prevalent amongst heather honey producers that, to obtain the very best results, it is necessary to re-queen each colony some time in July. However, I am led to the conclusion that it is a mistake to re-queen previous to the heather-flow — except in cases where bees are within reach of the moor from their permanent stands.   Re-queening will be found beneficial then, because a young queen will keep up a brood-nest of moderate size, whereas an old queen all too frequently ceases laying immediately the heather commences to yield.   But the position is quite the reverse where stocks are transported perhaps many miles to a distant moor. The jolting and general excitement caused by a journey rouses the bees to a renewed spurt of brood-rearing, to which one-year-old queens, still in their prime, respond more readily than queens only a few weeks old. We find colonies headed by one-year-old queens return from the moor in better condition for winter than either stocks which were re-queened prior to being taken to the heather, or those left on their permanent stands.

To introduce a queen of necessity entails a break in brood-rearing. A week usually elapses before a new queen is accepted and laying. Then a further two or three weeks pass by before a young queen is laying normally, and, moreover, a queen the summer she is born will never lay as well as when in her prime the year after. This, I believe, is a point not always fully appreciated.

Heather Hives

The selection of a suitable type of hive, for the production of heather honey, is an item of no small importance. A heather
hive must comply to the following requirements:

(1) It must be adapted for securing the hive parts and for confining the bees efficiently, with the minimum expenditure of time and labour.

(2) It must provide the strongest colony with ample ventilation and clustering space.

(3) It must be of a size and shape to permit loading closely together on a lorry or railway truck.

(4) It must provide adequate protection to a colony against the abnormal fluctuations in temperature which usually prevail on the moor.

Most appliance dealers sell hives which are specially designed for taking bees to the moors. It would seem, however, that in the majority of patterns insufficient attention is paid to the provision of both clustering space and ventilation. The single-walled American type of hive is admirably adapted for transportation, but hardly affords adequate protection in northern parts of Great Britain. The conventional WBC hive is perfect from the point of view of protection, but entails too much extra work in transportation, where a large number of colonies are taken to the heather. For many years we successfully used the internal fittings of WBC hives as temporary heather hives.

A floor-board was made for this purpose the exact outside width of a ten-frame brood-chamber, but extending three inches beyond the front to form an alighting-board. The day previous to taking the stocks to the moor, the roof and outer casings of the WBC hives were removed, the brood-chamber lifted, and the temporary floor-board placed underneath it. A shallow-super, or section-rack, was then put in position over the brood-combs, and on top of the super was put a wire-gauze super clearer to provide the necessary ventilation. The separate parts composing the hive were fastened together at each corner with crate staples.

On the morning after, about five o'clock, a perforated zinc strip was slipped into the groove at either side of the doorway and secured into position along the edge of the brood-chamber with tacks. The hive was then ready for the journey. Prepared in this manner, sixty hives, including roofs, can, with ease, be accommodated on a two-and-a-half-ton lorry.

The makeshift hive I have just described was used by us for ten years. We found it had one serious drawback. The standard ten-frame brood-chamber did not provide an adequate reserve of stores in seasons when cold, wet spells prevented the secretion of nectar until September. In such seasons, stocks were all too often on the verge of starvation by the time the weather improved.

Only one satisfactory remedy suggested itself, a larger brood-chamber—a brood-chamber commensurate in comb area to two ten-frame standard brood-chambers. But the general tendency in heather honey production was rather the other way, namely, for the reduction in the size of the brood-nest to endeavour to force the highly-prized heather honey into the supers. However, in our case the experimental hives, with brood-chambers of equal comb area to twenty British standard frames, in due course yielded most gratifying results on the moor. As I have already intimated, normally *no* undue amount of honey is stored in these large brood-chambers.

The brood-chamber of the hive we use is twenty inches square, and accommodates twelve frames of Modified-Dadant size. Whilst externally it resembles the American single-walled hive, it differs materially in many important details of design.

The floor-board has a slope of 1" from back to front, and the sides are rabbeted to prevent rain entering between the junction of the floor and brood-chamber. The entrance is 18" wide by 1" in depth, and may be reduced with an entrance block to 12" by 1/4"; the size of entrance invariably allowed on the moor. The floor-board extends only one inch beyond the front of the brood-chamber, and a detachable alighting-board is fitted instead.

Supers, both extracting and section-racks, are 6" deep, and accommodate either ten extracting combs, or alternatively, two extracting combs and eight section frames, each holding four 4 x 5 no bee-way sections.

A crown-board is used instead of the usual quilts, on top of the frames. The crown-board has an opening in the centre to
admit a double-exit bee escape to enable it to be used as a super clearer when required.

The roof is gabled, but constructed in such a way as to permit them being stacked on top of each other when they are removed from the hive for transportation.

The method employed for fastening the hive parts together for transportation to the moors is extremely simple, but nevertheless most efficient, expeditious and, moreover, absolutely safe. Two 1/4" diameter iron rods, threaded Whitworth bolt thread on one end, and a wing nut fixed stationary to the other end, form the whole outfit for fastening a hive together.

It is accomplished in the following manner: on the front and back of the screen are attached thin, small iron plates, which act as washers. In the centre of each of these plates is a 3/8" diameter hole, by which the iron rods are inserted and pushed down between the inside of the hive wall and the end bars of the frames, and through corresponding 3/8" holes in the bottom-board, underneath which are attached, with screws, iron plates, 3* x 7/8" x 3/8" in size, in each of which is a hole tapped to match the thread on the 1/4" rods. The rod, held by the wing nut, is screwed into the plate under the bottom-board, and the screen, shallow-super, brood-chamber and bottom-board are thereby securely and firmly held together.

Needless to mention, this method of preparing hives for transportation to the moor can be adapted to almost any type of hive. A blacksmith can provide the iron rods and plates for a few pence. As a matter of fact, the plates are not essential, nuts can be attached to the rods under the bottom-board instead. The plates are fitted for convenience sake and the saving of time.

When the bees are taken to the moor, one shallow-super is left on each hive to provide the necessary clustering space; but when brought back, the brood-chamber by itself furnishes enough room for the bees. Two lengths of rods are therefore required—one set for hives with supers, and a second set for bringing the hives back from the moor without supers.

The actual preparations for conveying bees to the heather are made some time during the day before transportation takes place. The screens and rods are placed in position, but the rod next to the entrance is not screwed tight.

About five o'clock the following morning the bees are first confined to the hives. This is effected by inserting a piece of wood into the aperture of the entrance contracting block. The rod next to the entrance is subsequently screwed tight; the piece of wood confining the bees, as well as the entrance contracting block, are thereby securely held in position. The hives are then ready for loading.

On reaching the destination on the moor unloading proceeds immediately. When the unloading is over, the iron rods are unscrewed, the alighting-boards placed into position, and then the bees liberated. A few good puffs of smoke are sent in by the entrance on withdrawing the piece of wood confining the bees, and also through the top when the screen is taken off.

The whole operation of loading forty hives, then unloading them on the moor and liberating the bees occupies about an hour and a half—this, of course, does not include the time spent on the road.

Anyone contemplating transporting bees to the heather might well adopt as his motto "Take no risks." To attempt to shift

bees in unsuitable or decrepit hives, or inefficient vehicles, or with unreliable assistance is indeed courting disaster.

Beekeepers are optimists, and it seems, therefore, their fate to learn in the school of experience.

Brother Adam with Colin Weightman

Drifting

I believe text-books tell us to arrange hives with entrances facing south-east. This, no doubt, has given rise to what now

seems a universal custom of placing hives, in straight, symmetrical rows. As a matter of fact, there was no other alternative if entrances must face south-east. However, is it the most satisfactory arrangement from the practical aspect of the question? Indeed not! Anyhow, not on the moor.

If hives are placed in symmetrical order, all facing one direction, then the colonies towards the end of each row, nearest to the moor, possess all the field bees at the close of the season. Drifting on the moor is a great problem. When bees return from the heather laden, they tend to enter the first hive in their line of flight. When heavily laden with nectar, they are freely admitted by any colony. One is then inclined to argue: "What does it matter, the honey will be obtained nevertheless." Unfortunately, things do not work out that way. It will be found that these abnormally strong stocks never store honey in amounts corresponding to their strength. Furthermore, after the honey-flow, constant lighting takes place, and not infrequently queens are, in consequence, balled and killed. We therefore endeavour to check drifting by every means at our disposal.

The most effective way to avoid drifting is to arrange hives in groups—four hives to a group, and the entrance of each is

made to face a different point of the compass. Furthermore, if posiible, each group is placed next to a distinguishing mark, as, for instance, a gorse bush, a clump of ferns or a boulder.

There is a further, perhaps the most serious aspect respecting drifting, namely, the domination of disease. If there is a diseased colony present, drifting will speedily transmit the malady to other stocks. Indeed, do we not often hear reports of stocks returning from the moor diseased which were perfectly healthy before? I know of a commercial beekeeper who transports hundreds of hives to a moor a hundred and forty-five miles away rather than place them on a nearby heath, where foul-brood is known to exist.

18

Shelter

Shelter from prevailing wind is another important detail conducive to success in the production of heather honey. Where natural shelter is not available, it is well worth the trouble and expense of erecting artificial wind-breaks.

The lee-side of a hedge forms an ideal place for an apiary.

The inestimable value of shelter was demonstrated to me many years ago, in a most forcible manner. A short distance from an apiary which was situated in a gully were twenty-eight hives which, however, we had been compelled to place out on the open moor. One day, when visiting both places, the bees in the gully were found building comb and storing honey quite actively, whereas those in the bleak situation were on the verge of starvation.

The consideration Dartmoor farmers evince respecting the provision of shelter for the bees has always greatly impressed me. These hardy moor-men are most solicitous that the bees should have the advantage of what they call "a looe place."

Dartmoor

Before proceeding with the subject I feel sure it will not be amiss to give a brief description of Dartmoor for, it must be
remembered, much what is said relates mainly to conditions as found on Dartmoor.

It may be recalled that the Celtic name for Devon was Day-thaint, which means "dark and deep valleys." The south-east approach to Dartmoor is by way of dark and deep valleys. The Dart, the most romantic of rivers, winds its way along in such a valley.

The outskirts of the moor are mostly densely wooded but the moor itself, although spoken of as " the forest", is devoid of trees. Dartmoor extends thirty-two miles trom south to north, twenty-two miles from east to west, and comprises an area of 130,000 acres. It attains a height of 2,039 feet above sea-level. The forest, except for the Bell of Commons which surrounds it, is the property of the Prince of Wales. The bordering parishes exercise over commons, and to some extent over the forest, rights known as " Venville." Members of venville parishes claim the prerogative to take from forest and commons "anything that may do them good" except green-wood and venison. Tenants in venville pasture their flocks and herds, cut turf and take such stone and sand as they require.

Dartmoor is of volcanic origin, and is mainly composed of granite. The granite, however, is not of uniform quality. Apparently it was not well stirred before being poured out, hence weathers most irregularly! The wreckage, locally termed 'clitter', thus formed of the tors, in the course of time in some cases covers the entire hillsides.

The moor is divided in halves by a road traversing it from east to west.

Near its western extremity is situated Princetown Prison. On the northern part of the moor, neither road nor any human habitations exist.

On the southern half, along the central road, are here and there, surrounded by small plots of cultivated land, a few ancient tenements—the only farms extant on the forest. These farms date back to remote antiquity.

The rest of the moor is a vast, inhospitable wilderness. Long ridges rise behind the other, like waves of the sea. However, the wide, solitary wastes, the rugged, rock-strewn peaks and giant tors impart to Dartmoor a peculiar fascination.

But it is a land of silence—a silence that is awe-inspiring. The faint rustling of cotton-grass and heather, the hum of bees and purling of the moorland streams are the only sounds in this solitude to strike the ear.

Yet in prehistoric times the forest was densely populated. Everywhere are remains of the ancient moor-dwellers. Cyclo- pean bridges, trackways, stone avenues and hut-circles the ancient Celts built and used can still be seen.

However, it is the honey and pollen-bearing flora of the moor we are foremost interested in.

In spring gorse and broom provide pollen, bluebells and whortleberries nectar. Occasionally gorse flowers a second time simultaneously with the heather.

Ericacea

The family of heaths, which botanists call the Ericaceae, include an amazing variety of shrubs and trees. Beekeepers are primarily interested in the genus *Caliuna vulgaris*, and in the two common Ericas, *cinerea* and *tetralix*; and to a lesser extent in the three rarer species, *ciliaris, vagans* and *cornea*.

The Dorset heath, *E. ciliaris*, the Cornish heath, *E vagans*, and the Irish heath, *E. cornea*, are only found in the south of England and Ireland respectively. They are of little value to bees, and we shall therefore not consider them further.

Calluna vulgaris, or ling, *E. cinerea*, or fine-leaved heath, *E. tetralix*, or cross-leaved heath, are found on moors and forests throughout Great Britain.

Erica tetralix. commonly known as bog heather, is usually only found on wet, boggy parts of the moors. It is short in growth and very bushy at the base. The leaves are in fours, formed at each joint of the stems; this fact being set forth in its title, "tetralix" meaning four-leaved. It is also called cross-leaved heath, for when viewed from above, the four leaves assume the form of a cross. It blooms in July. The flowers are most gracefully formed, and the wax-tike, rosy-hued bells are always borne at the top of the spikes, in drooping little clusters of close umbels, all facing in one direction.

Bog heath, though not as showy as bell heather, is undoubtedly the most charming of the two. It is not abundant on Dartmoor, and generally grows only to a height of a few inches. I know of only one spot where it develops to perfection.

E. tetralix yields pollen and nectar. *Erica cinerea*, commonly called *bell heather*, forms a bushy growth. Its branches spring mostly separate from the base, and grow to a height of about twelve inches. The leaves are sharply pointed, and grow in whorls around the stems. The flowers are a rich, reddish purple, produced on upright spikes in dense terminal racemes one above another, and are bell-shaped in form. Bell heather is in full bloom about mid-July, but it can still be found in flower here and there throughout the autumn.

E. cinerea flourishes mostly on the foot-hills and outskirts of Dartmoor. It does not thrive on the higher regions of the moor. Bell heather is the main source of nectar on the low-lying moors of southern England. On Dartmoor, on the contrary, I have known it to yield heavily only one season, in 1920. The honey derived from this source is usually a lowly crimson-red in colour, but in some districts is as dark as treacle. It is fair in flavour, but lacks body, and granulates with a course grain.

Calluna vulgaris. or *ling*; whilst not so attractive as the Ericas, is nevertheless of foremost economic value. It is from the ling that the world-famed, the true heather honey is derived.The ling forms a rather straggly shrub, usually about a foot in height. Its branches are tough and more woody than in the other two species. The leaves, borne on close masses on the side shoots, are very small and, moreover, are covered with knobby hairs. The flowers vary in colour from a deep purple to a very pale pink. and occasionally are found pure white.

The ling is in flower from July 25th to September 20th in normal seasons, but secretes nectar generally only from mid- August until September 5th, the period when it is in full bloom.The honey varies in colour from a light amber to red-brown, has a jelly-like consistency, a bitter-sweet flavour, and an all-pervading fragrance.

A complexity of factors, of which as yet very little is known definitely, affect the secretion, flavour, colour, viscocity, and, to some extent, aroma of heather honey.

Subsoil

The mere presence of an abundance of bloom does not denote that nectar witi be forthcoming. Very little honey is derived from ling, for instance, on the moors around Aldershot, the New Forest and in Dorsetshire, although there is a wealth of bloom, in fact far more of it than anywhere on Dartmoor. Whilst a lime-free surface soil is essential for the well-being of heather, the absence of nectar is accounted for by the particular subsoil of the southern moors. I believe it is generally maintained that no nectar is secreted on sandy or chalky subsoil, and that ling yields most freely, and gives honey of prime quality, on granite or ironstone —which our own experience seems to corroborate. Whereas almost the whole of Dartmoor is granite, the heather is singularly confined to the area where tin-streaming was in progress in centuries gone by. Is the presence of iron in the tin-bearing ore the cause of this curious phenomenon?

Swaling
Another condition affecting the secretion of nectar is the formation of growth, or the age of the plant. When ling is permitted to grow on for years, without any interference, it forms rank, woody growth. Although it will then still flower, the nectar obtained therefrom is quantitatively less, and also considered qualitatively inferior.

The burning-off of heather, or "swaling" as it is termed by moorfolk, is therefore a blessing in disguise to beekeepers. This we never fully appreciated until the past dry summer. The parts of the moor which were burnt over two or three years ago produced this season a veritable mass of bloom, whereas the old, straggly heather, acres and acres of it, hardly formed any bloom. A Staffordshire beekeeper reported the same occurrence on Cannock Chase moors this season.

Subsequent to a conflagration, two or three years have to elapse before the heather comes into full bloom again.

Altitude
Originally, our apiaries were situated on the southern outskirts of Dartmoor, at elevations of approximately nine hundred feet above sea-level. The locations were considered ideal for the production of heather honey. However, a few years later an apiary was established at an altitude of 1,500 ft. The difference in the quantity and quality of the honey obtained on the higher moor was most remarkable. Now nearly all our apiaries are located around the highest points, approachable by road, of the southern half of Dartmoor. Last year we secured about 32 lb. more honey from colonies at elevations of 1,500 ft. compared to the stocks still on the lower ranges of the moor. As conditions were identical in all localities, where the apiaries are situated, altitude, in this case, must mainly account for the difference in the yield. Still there is no denying the fact that heather honey is obtained at elevations below 800 ft. But to get heather honey of supreme quality, a high altitude seems essential.

Climatic Conditions
The most outstanding characteristic of ling is its ability of yielding in minimum temperatures, when no nectar could reasonably be expected of any other flowers. Instances are on record when it actually secreted heavily subsequent to a severe hoar frost. I remember one season, 1923, when the

maxima never exceeded 67 degrees F throughout the duration of the flow. We nevertheless obtained a crop which in quantity, and especially in quality, left little to desire. Notwithstanding, nectar is secreted most liberally with night temperatures of not less than 42 degrees F, and day temperatures fluctuating between 70-78 degrees. But, so I am led to conclude, a humid atmosphere, with a drift of air from a southerly direction, is as essential for a heavy flow as any other factor. With a current of air from the north-east or north, there may be every other condition present conducive to a flow, but no nectar wtll be forthcoming. The ideal close, sultry weather seems invariably of short duration at the time the heather is in bloom. After three or four perfect days for nectar secretion, thundery weather develops, which generally denotes the conclusion of the honey-flow that season.

Duration of Honey-Flows
Seldom or hardly ever is a heather honey-flow protracted. In most seasons it lasts but a few days. In 1917 heather honey was gathered only during a couple of hours one afternoon. The most protracted flow I remember extended over a period of twenty-six days without any cessation. Although nectar is secreted on the moor during comparatively brief spells, this is counterbalanced by the fact that no other honey plant in the British Isles yields as profusely when conditions are favourable. Instances are on record where stocks of average strength made gains of 10 lb. in the space of a *few* hours. The yield from August 25th to the 28th last year was undoubtedly the most phenomenal flow we have ever witnessed.

In normal seasons, no heather honey seems ever to be gathered either before mid-August or after September 10th. So on or about September 5th all the supers are placed on bee escapes. Thereafter we endeavour to lose no time in getting the hives back from the moor, to permit wintering preparations to be completed by October 1st.

Feeding
This brings me to the much debated question of the suitability of heather honey as a winter food for bees.

Many experienced heather men declare it absolutely detrimental, whilst others, on the contrary, hold heather honey to be an excellent winter food. Our own experience in this matter leaves no doubt that heather honey cannot, year after year, be relied upon to carry bees safely through winter. We have therefore made it a practice to feed each colony about 15 lb. of syrup, regardless of the amount of stores they may have before closing down for winter. Ever since, our winter losses have been negligible. The syrup fed is composed of sugar and water only; nothing else is added. The proportions are: 7 1/2 lb. water to 10 lb. sugar. The syrup is prepared by the cold process, that is, the sugar is merely dissolved in cold water.

Our tank was specially erected to facilitate the preparation of syrup by the cold process. The correct quantity of cold water is allowed to flow into the tank, then the sugar added, and an occasional stirring to keep the sugar in suspension is all that is required to dissolve it. A ton of sugar takes about two hours to convert into syrup.

The tank is lined with glazed tiles, with the object of avoiding every possible danger of contamination, as would be the case with a metallic container. We are, moreover, thus enabled to prepare the syrup long in advance at our entire convenience.

Extracting

Whilst the work of bringing back the hives from the moor, and the feeding and preparing for winter are in progress, the extracting of the heather honey is also proceeded with. Immediately the crop is off the hives we endeavour to extract it without delay.

Honey derived from bell heather can be extracted by means of centrifugal force; in fact, it leaves the combs more readily than flower honey. The true heather honey—the honey of jelly-like consistency—no amount of centrifugal force can possibly dislodge from the comb, unless the contents of each cell are first loosened and then extracted before the honey re-assumes the gelatinous state. This, in fact, is the principle involved in the heather honey loosening machine, first placed on the market in 1906. But as bees build comb irregularly, it is next to impossible to construct a machine which will effectively loosen the contents of each cell of a comb. Moreover, even after treatment with a heather honey loosener, genuine ling honey will not extract as cleanly as flower honey, and unless the extracted combs can be given back to the bees to be cleaned before cold weather sets in, the remaining honey will deteriorate, thereby affecting the crop subsequently stored in them. Also, in the process of treating heather honey, the needles of the honey loosener unavoidably break down some cell-walls which during extracting are bound to get into the honey. Due to the peculiar consistency of heather honey, it is practically impossible to free it of the particles of wax subsequently. There is, to my knowledge, no satisfactory means by which ling honey can be strained.

Where heather honey in run-form is produced, a press of some kind is essential. If only a small quantity of surplus has to be dealt with a hand squeezer, made of two pieces of wood each two feet in length and three inches in width by an inch in thickness, hinged together at one end, will answer the purpose. The honeycomb is cut out of the frame, tied in cheese-straining cloth and the bag hung up. Then the comb is crushed with the hand squeezer, and thereby the honey is liberated from the wax. We ourselves used an appliance of this kind many years ago. However, where a large crop has to be dealt with, this method is too wasteful and slow so a specially constructed honey press is then required.

A press must be substantially constructed to extract honey efficiently. Very considerable pressure is demanded to get honey satisfactorily out of comb. We have used no less than six different types of presses. Eventually, as there is no press on the market for dealing with large crops efficiently and expeditiously, we were compelled to design and construct one of our own.

To operate the machine, the pressing-board is first detached from the beam, and pushed back out of the way on the trolley. A sheet of cheese-straining cloth, six feet square, is next spread over the grid container. The slightly warmed combs of heather honey are then cut from their frames and laid side by side and five high on top of each other inside the press container, which admits fifty shallow-combs at one pressing. When full, the overlapping straining cloth is folded over the top of the combs, the pressing-board pulled forward into position and attached to the beam; the beam, with the pressing-board suspended on it, is then slightly raised to permit the trolley to be pushed behind the press. The motor is then stopped and switched into reverse, when slowly but surely the pressing-board is forced down on the honey-combs. Immediately the motor indicates stress, the low gear is engaged, which transmits twelve revolutions to the worm-drive, and a pressure of over a hundred tons to the square inch develops. An idea can be formed of the efficiency of the machine by the fragment of a sheet of wax taken from the press. From the six inches of combs a sheet of nearly solid wax, 3/8" in thickness, is all that remains inside the straining cloth. The marks of the wire screen, impressed on the wax, are plainly visible. After each pressing the cheese-straining cloth is rinsed in clean water to prevent the bits of wax adhering to it from getting into the honey already extracted,

An almost insuperable difficulty, connected with every type of honey press we used, was to find a cheese-straining cloth which would withstand the enormous stress it had to be subjected to in pressing honey. The straining cloth supplied by appliance dealers was, after two or three pressings, torn beyond repair. A year ago we were eventually successful in procuring a cheese-straining cloth of the quality required. One sheet of the new straining cloth enabled us to do no less than seventy-five pressings before it had finally to be discarded.

As already stated, the honey, after it has been extracted, collects in a tank at the rear of the press. From there it is delivered by a specially designed power-driven rotary pump into the storage vats on the second floor.

Whilst seasons of plenty and scarcity do not come in periods of seven years, nevertheless, to assure customers a continuous supply it was found imperative to provide for seasons of failure, which, especially in heather honey production, greatly predominate. With this object in view special storage vats were built.

There are eleven tanks, totalling a storage capacity of nearly thirty tons.

The internal dimensions of each tank are 7' x 3' x 3'. They are lined with tinned steel plate, made to our requirements. The lids, on account of condensation, are covered on the underside with aluminium. Between the lids and the top of the tanks is a rubber tape which, when the lid is bolted down, seals the tanks hermetically. The honey can therefore be kept indefinitely, without any deterioration, in these tanks.

Filling Machine

We use an electrically operated bottle-filler. This filling machine enables a tank full of honey to be emptied into I lb. containers in the space of three hours, or it fills on an average 2,000 cartons in sixty minutes. Each container is filled to the exact weight, without any drips.

In each of the eleven tanks is installed a coil, composed of 125 ft. of 1" bore tinned copper tubing. The coils are connected to a gas-heated, thermostatically controlled boiler. The temperature of the water circulating in the coils can be regulated to a nicety. When the coils are empty, they can be detached and bodily lifted out of the tanks for cleaning. The time required to liquefy a tank full of solid, granulated clover honey, two and a half tons of it, is eighteen to twenty-four hours.

This brings me to the question of granulation of heather honey, or, to be more precise, granulation in heather honey. For, I believe, it is generally admitted that pure ling honey never granulates, which our own experience seems to confirm. In 1923 no heather honey was gathered on Dartmoor until September 8th and then throughout the duration of the flow, the maximum temperature was so low that, even if there had been other flowers, no honey would have been secreted by them. In consequence, the crop that year was probably pure ling honey. Anyhow, it never manifested the slightest sign of granulation. Occasionally the honey we obtain granulates to a consistency of soft butter, but the typical Dartmoor heather honey stays liquid; crystals are formed only here and there, kept suspended in the honey. This kind of granulation is, no doubt, accounted for by a slight admixture of honey derived from bell heather.

When for any reason heat has to be applied to heather honey, the utmost care has to be exercised. In no case must the temperature of the honey exceed 130 degrees, or be kept at this temperature for any length of time. There is no other honey of which the flavour, aroma and consistency is so easily ruined by the application of heat. Heather honey swells considerably when heated, and allowance must therefore be made for the expansion which takes place.

Heather Honey

We have in our possession a collection of samples of heather honey derived from all over the British Isles. Hardly two samples are alike regarding colour, flavour or granulation, and, although considered pure heather honey, some flow exactly as ordinary flower honey does. Apparently, any honey obtained on or near a moor is designated "heather honey". However, there is a decided and unmistakable difference between the true heather derived from the ling, and bell heather, or the numerous blends. Ling honey is distinguishable from other honeys by a simple test. If, on inverting a jar of liquid honey, the contents do not flow, then it certainly is heather honey. I do not mean to imply that is 100% pure, but that true heather honey predominates.

Although Ericas are commonly spoken of as heather, they are to be precise, heaths; the ling, heather. Heaths and heather belong to the same family, but of the ling only one species exists, whereas of the genus Erica there are numerous species, five of which are indigenous to the British Isles. Moreover, the true heather honey possesses a flavour, aroma and consistency absolutely unlike any other honey. That from the heaths has not the faintest resemblance to it, excepting colour.

The Ministry of Agriculture has recently issued regulations whereby heather honey is now placed under the National Mark. This undoubtedly is a great step forward in bringing heather honey into its rightful position.

All our heather honey is marketed in a special carton.

(After the meeting Brother Adam was subjected to the usual bombardment of questions, which he answered with great courtesy. A very hearty vote of thanks was passed, but the most complimentary token of regard to his great work in making beekeeping a real business proposition was the exceptional gathering of nearly three hundred beekeepers from a large area of the north).

WILLIAM WILSON SMITH
1890 – 1969
BY MORNA STOAKLEY
(PEEBLESSHIRE BKA)

Willie Smith with his hives fastened together ready for transportation.

●

(first published in the Scottish Beekeeper, September 2012, and reproduced by kind permission of the author and editor).

Willie Smith of Innerleithen is, without a doubt, Peeblesshire BKA's most famous member as well as being one of Scotland's first successful commercial beekeepers. He was born on 25th February, 1890, at Gatehouse, St. Mary's Mount, Peebles and baptized in Manor Kirk in April of that year. After service as a despatch rider in the First World War, he was employed as a chauffeur by Mr. Ballantyne, a well-known mill owner in Walkerburn. During his spare time he took up beekeeping. His hobby became a very successful sideline and, encouraged by Mr. Ballantyne, he slowly built up his colonies. In 1933 he was running about 50 colonies and by 1945 he had expanded to 150 colonies and was obviously making a good living as he was able to run his own motor car – quite unusual at that time. The number plate of his car was DS 1819; by 1950 he was also running an old blue van – DS 1879.

All this time Willie had been experimenting with various different kinds of hive. He was certainly using the CDB (Congested Districts Board) hive. We think he may first have come across the CDB hives when he was posted to Ireland in 1916 - the Irish Government at that period developed and funded the hives and ran many courses for training beekeepers in a bid to help reduce rural poverty and provide additional income sources for country people. The CD Boards apparently had some involvement in Scotland too. Double-walled hives, such as the WBC, are cumbersome and time-consuming to use and very difficult to move to other apiaries, particularly moorland heather sites. The National, or Modified National, is very fiddly to construct – though the side battens (put on to accommodate the long lugs of the frames) do make it much easier to lift heavy boxes compared to the narrow indents on Smith and Langstroth hives!

Willie had also tried the American Langstroth hive and while he liked its simplicity of construction he found the boxes generally too large for the Scottish climate. So he set about making a simple Langstroth-style box of four sides with rebates but to fit a British Standard frame with the lugs shortened. He also chose to make it with top bee space as this allowed him more easily to tilt up an upper brood box to make a quick check for queen cells. Willie liked to work double brood boxes during the early part of the summer and then reduce down to one brood box to give him very strong colonies for the heather – a potential major crop in this part of the world. Willie had originally put plinths around the bottom of his boxes but after importing the Langstroth hives he realised it wasn't necessary.

His hives had a top bee space which made it an easy matter for checking for queen cells when two brood boxes were used.

Willie became a very active member of Peeblesshire BKA. He was Secretary and Treasurer from 1924 until 1935 and then was elected President from 1939 until he stood down in 1945. The Minutes of PBKA for June 1941 state:

Mr. Smith was called upon to proceed with his demonstration, during the course of which he displayed a unique hive which he himself designed, the main feature of which was the ease by which it could be manipulated. He advocated that appliances should be standardised (as in the United States and Canada) as there were so many hives advertised by rival manufacturers so elaborate as likely to confuse the novice, as after many years experience of beekeeping on a large scale he had decided on simplicity.

In 1942 Neil Anderson from Fife spent some time working with Willie Smith and took back drawings for what Willie called "My hive". Once back in Fife Neil drew up detailed plans for this hive and called it the "Smith hive". He sold these plans for a "modest cost" and eventually the British Standards Institute sent for a copy with a view to drawing up an official Standard for the hive.

There are several references in PBKA Minutes to the progress of the naming and standardisation of Willie's hive – and the SBA is reported to have not been in favour of a Surname being attached to it! The PBKA Annual Report for 1947 states:

Regarding the Smith Hive, the following resolution, proposed by Mr. S.L. Shannon and seconded by Mr. Alan Smith, was passed at the meeting of 26th October 1946: "In view of the fact that the Specifications of the National Minor Beehive as published by the British Standards Institution are practically those of the 'Smith' Hive, this Association wishes to protest against the injustice done to Mr. Wm. W. Smith in adopting his hive with slight modifications and giving it another name. Further this Association is of [the] opinion that any departure from the specifications on the 'Smith' hive would be against the best interests of beekeeping."

Then further on this same Annual Report tells us:

At the Annual Summer Demonstration on 7th June, 1947, Mr. John Murray, President, introduced Mr. Wm. W. Smith and congratulated him on behalf of the Association in having his type hive accepted by the British Standards Institution.

The original blueprint for the Smith Hive is still in the possession of his granddaughter in Innerleithen.

The Minutes also tell us of a visiting speaker to PBKA in January 1948:

... Mr. Charles Bruce, Secretary of the East Lothian Association, an organisation with one hundred and eighty members ... is a schoolmaster in Pencaitland and instructs a woodwork class ... he finds that his pupils can make two Smith hives in about the same time as one National.

I don't think anyone who has made both kinds of hive would dispute that!!

Willie Smith was a member of the Honey Producers' Association (now called the Bee Farmers' Association) for many years. Willie Robson (of Chain Bridge Honey Farm) recalls a visit made by R.O.B. Manley and A.W. Gale, both well known commercial beekeepers of that era. Selby Robson (late father of Willie and a former College Beekeeping Advisor) had met them off a train and taken them to Willie Smith's apiary.

...When they arrived at the apiary they were attacked by bees and Willie could not be found. Needless to say he had retreated into the bushes. In days gone by when protective gear was rudimentary in the extreme it was very common for beekeepers to retreat into the undergrowth to get rid of followers. On that day the bees were not out of hand, just extremely peppery. This caused a good deal of amusement...

Willie Smith's apiary at Kirkhouse.

When Willie gave up beekeeping he passed most of his equipment and bees to George Lunn, an apiarist at the College (Edinburgh) who had frequently helped him. He also passed some on to George Hood of Ormiston – who continued to use the Peebles area for heather honey until his untimely death in 2010.

After Willie's death in 1969 the Smith family presented the SBA with the 'W.W. Smith Memorial Trophy for exhibitor with most points in the heather honey classes at the Scottish National Honey Show'. For many years this trophy was won by Bill Foubister from Alford. More recently it was frequently won by our own Jim Bogle with good Peeblesshire Heather Honey!

We have made digital copies of a great many photographs lent to us by Willie Smith's granddaughter and also of the original Blueprint. It is amazing to see folks walking around apiaries all without protective clothing – and often wearing their Sunday best! We shall be circulating some of these photos round older SBA members to see if we can identify some of the people in them.

Although it is sometimes now replaced by polystyrene hives, I am sure that the Smith Hive will continue in use as long as there are beekeepers in Scotland! Sadly, Willie died the year before John and I moved to Peebles so we never met him.

Very many thanks are due to Mandy Clydesdale (current Secretary of PBKA) and Jean Dobie (Willie Smith's granddaughter) for much help in sourcing material.

BEES AND LEY LINES!

Bill Clark

On Saturday July the 14th 2012, my memory was jogged about an article entitled: Bees and Ley Lines - in the August edition of the 2011 BKA News, to which I had meant to respond. Yes, I am afraid I have reached that certain age! What jogged my memory? A group of folk outside our kitchen window, being instructed in the joys of dowsing. In this case they were trying to map the rooms within the foundations of a long-demolished mansion. But I am sure if I had followed when they moved off to investigate our alleged Chalk Figure, as on other occasions I would have heard Ley Lines mentioned.

For a full history - and this has nothing to do with my age - I need to go back to my childhood. I first tried to dowse in the company of my grandfather, who was using a forked hazel stick to look for a lost well. I was useless. Later, I saw a professional water diviner at work. "You should use a willow stick." I was still useless! Then in the 1950s, whilst I was working with earth moving equipment, an engineer showed me how to find an underground cable, using a couple of brass welding rods, bent in an L shape. From then on I regularly used them - it was the best grounding - no pun intended - in dowsing that anyone could wish for.

At first, I only looked for electric cables - alive or dead - iron, copper, or lead pipes, house drains and clay field drains. If any were water-filled I got

33

an especially good 'kick' which led me to believe that I could find water too, and did! A good party trick was to pass them over a glass of beer, and see them clash together. Because I usually uncovered my finds later, I knew for certain what lay below. Once, when excavating low lying farmland, ready for a factory to be built, I was puzzled to find nothing where my rods had strongly indicated, until I realised that there were lines of different soil colour, filled in trenches - and I remembered grandfather describing how they used to drain fields with 'bush drains' by burying the hedge trimmings in the bottom of trenches. As I progressed, looking at my surroundings first often gave me strong clues, even before I used the rods.

For some ten years, I had no reason to use the rods, until I arrived to be in charge of this estate in 1973 and discovered a smelly wet patch out in the grounds. Obviously a blocked drain, but from hence, and to whither? With my brass rods long lost, I made them from the only thing to hand - a length of 5 mm copper brake pipe. My eight-year-old daughter, Caroline, seeing me testing the rods, asked to try. I steered her to walk over the water pipe to our house - over a metre deep. "Ow," she exclaimed, throwing down the rods, "A shock went right up my arm!" Needless to say, I tracked the drain in its entirety, and after putting it to rights, went on to track the other strong lines around the remaining buildings and garden area. Manhole covers along the way usually told me when drains were below, but others needed a small 'pilot hole' digging, in order to ascertain the identity. I finally knew the extent of the foundations of other long-gone buildings, buried gravel paths, early brick drains and lead water pipes. Even the later galvanised iron ones, running out to both vanished and existing horse troughs in the fields. Unfortunately, spread all over the area too is the activity of Bronze and Iron Age villagers, my pilot-holes were taboo in these protected areas, but I was well satisfied when many of my 'hits' turned out to be finds during 1990s Cambridge University archaeological surveys!

It wasn't until 1974, that I realised archaeologist T C Lethbridge, who had carved out a controversial chalk figure on the estate during the 1950s, was causing us to be a magnet for dowsing enthusiasts determined to find out for themselves - is it or isn't it? A popular cry was, and still is, "Is it on a Ley Line?" The most excited folk find a number of Ley Lines heading straight for it! But once bitten by the bug, they usually move from checking other's work to finding their own Ley Lines. Strangely, although Lethbridge himself had long been a hazel twig dowser, looking for ancient graves and such, later writing about the use of the pendulum and paranormal activities, he didn't use these methods to locate his figure.

What are Ley Lines? They are believed to be geopathic stress lines that emit energy. I can quite believe that there are energies emanating from the earth's magnetic field and probably also from tectonic plate and earthquake movements. But I have only seen these folk following my known lines of 'force.' I had to bite my tongue as, compass and map in hand, one said to a companion - marking along a particular drain - "This Ley Line is heading straight for Kings College Chapel." On another occasion, a professed expert told me, "This Ley Line links up with Wimpole Hall." I had no idea our old water pipe went that far! The zenith has to be the gentleman who, after months of research said, "I can now reveal that Wandlebury is the epicentre for most of the Ley Lines in Britain". There was a hole in the centre of Wandlebury Ring on his map, caused by the number of times his fountain pen had crossed through! One lady with a house and garden in the Cambridge suburbs had two parallel Ley Lines running through it. A glance at the map showed the house had been built on a closed, 'Beeching line." I could easily locate the two lines of rust from the rails, and metal from clattering train wheels that a century of rain had washed down into the soil. Probably my finest hour was seeing a JCB digging a trench alongside a country road, about to cross a line between two sets of distant, 30,000 volt pylons - both had thick cables running down to the ground. My rods nearly jumped out of my hands, just one metre short of the incredulous driver's trench.

So, for the disbelievers, I should have successfully buried Ley Lines! But believers will still argue that most of the old tracks, especially Roman roads, are on a Ley Line; I have proved that both dug ground and compacted ground give a signal. Ancient oak trees are found on Ley Lines, especially at junctions. I say they were often planted, both alongside and at track junctions. Particularly well grown oaks are often on a Ley Line. I have noticed oak trees grow well along spring lines - which give a very good dowsing signal. One field has many Ley Lines, whilst another none. Few grazing fields were drained, only the problem arable fields. I remember my father being told by an old man, "That field has never been drained." We found a network of three series: very early stone, early 19th C clay horseshoe, and mid-9th C handmade clay pipes. Ley Line enthusiasts would have had a heyday! I once located straight lines two metres apart over an entire field, the farmer was most impressed. He said steam engines had mole drained the field during his childhood. I am sorry, but I have yet to be shown a Ley Line that I haven't been able to figure out another very good reason to make the rods click together. One enthusiast for the hill figure once said, as we battled it out on the Cambridgeshire Radio, "It's all right for Bill, he will insist on dealing in facts."

Now, at last. Bees and Ley Lines! A barn that is a magnet for bees? The photo in the BBKA News which showed the type of barn that could have been built for bees to colonise. It only needs one poor beekeeper in the area to keep it going, and a swarmy season for the barn's own colonies to proliferate. Another construction, also designed to be a magnet, is the flat roof; fortunately, in a well-constructed flat roof there is seldom more than one hole, but in a barn like the one shown there will be dozens. I have often been called to both; such barns often had multiple bee and wasp colonies, the flat roofs only the one. Do swarms follow Ley Lines by choice? I know of three houses that are - according to our enthusiasts - on Ley Lines, and three hollow beech trees that I am sure they would say are not. All these locations have regular swarms entering. I once followed a swarm from my hill top to another - a definite candidate for a Ley Line. They landed on the tiles of a house hidden among trees and proceeded to run into the roof of a dormer window. The house owner was not at all surprised "As fast as I get them killed off, another lot arrives, all my dormers have had them in." I advised him to clean out every last vestige of wax and seal up every hole. He then, unwittingly, gave me the reason for his house being the chosen one. "Funnily enough, a beekeeper used to live here!"

Do colonies have fewer varroa when living on a Ley Line? I could certainly point out many reasons to consider first!
  (1) The beekeeper and his methods.
  (2) Nearby Beekeepers and their methods.
  (3) The nearness of other colonies - beekeepers or wild ones.
  (4) Are they healthy over two or three seasons?

As perhaps is becoming evident, even an enthusiastic (4) would not convince me that a Ley Line was responsible! In fact, given our Island's long history, and my proven causes for my dowsing rods to indicate, I would say that the beekeeper desiring NOT to put his hives on or near a 'Ley Line' would have the greatest difficulty!

ALFRED WATKINS
THE MAN BEHIND
THE METER
BY PETER TOMKINS

Alfred Watkins: A man of many talents.

●

(originally published in BKQ No 48, Winter 1996/97)

"He was scholar, miller, farmer, archaeologist, naturalist, inventor, magistrate, county councillor, politician and leader of public opinion. He was full of years and honours..."

Alfred Watkins' obituary in the Daily Express of the 9th April 1935 listed some of this remarkable man's accomplishments but failed to mention two of his interests which are of particular relevance to beekeepers. He was a founder member of the Hereford Beekeepers' Association and did much to promote the craft in that county,   and as an accomplished photographer he recorded a magnificent collection of rural scenes of Herefordshire many of which have a beekeeping theme. It is, though, for his controversial theory of 'Ley Lines' that he will be most remembered. Watkins demonstrated that in pre-Roman times ancient sites, hill forts, burial sites and mounds etc., were aligned with prominent natural features to mark out trackways; his book "The Old Straight Track" is still in print.

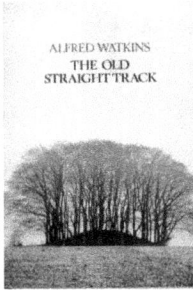

ALFRED WATKINS
THE OLD
STRAIGHT TRACK

Alfred was born and lived all his life in Hereford only making occasional sorties outside the county. Charles Watkins was a local innkeeper at the time of his son's birth in 1855 who later bought the principal brewery in Hereford and several other businesses in the area including the Imperial Flour Mills and the Imperial Hotel. It was while working as an 'outsider' at the brewery and later at the flour mill that Alfred travelled the length and breadth of the county, giving him an intimate knowledge of the people, places, and history of the area. He was to produce many articles and books on a variety of subjects, often illustrated with his own photographs; his notes on early pottery kilns are referred to at the present day.

In 1890 Watkins converted a room at the Imperial Mill into a workshop and it was here he developed and produced the Bee Meter referred to in earlier editions of The Beekeepers Quarterly (Nos.45 & 46). The meter's success was assured when it was used to help produce outstanding photographs by H.G. Pouting, the photographer who accompanied Scott's Antarctic Expedition in 1910, the same year

Watkins became a Fellow of the Royal Photographic Society and was awarded the prestigious Progress Medal for his research into photographic theory and practice. The Watkins Meter Company produced several other aids to photography and also developed a 'time and temperature' method of processing film that was still in use by the R.A.F. during World War II.

Advertisement for the Watkins' Bee Meter. First produced around 1900, these meters are frequently available on Ebay, often for well below £10.

It will be the beekeeping activities of Alfred Watkins that are of most interest to readers of the Annual. He was familiar with bees from a very early age; he recalled working with a Woodbury bar frame hive in the 1860s in a bee shed built by his father.

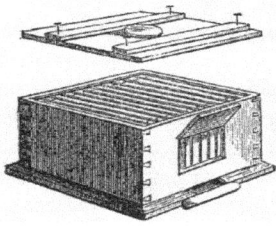

In 1882 he became a founder member of the Herefordshire Beekeepers' Association and was its secretary until 1901. Amongst the vast collection of his photographs held in Hereford City Library are many relating to beekeeping; they provide an invaluable record of the period from the 1880s until his death in 1935. At a time when skep beekeeping was prevalent and honey removal often involved killing the bees with a sulphur candle, Herefordshire County Council took the highly commendable step of funding a mobile Bee Van to tour the county and "teach good and efficient bee-keeping". An instructor travelled with the horse-drawn van and after a practical demonstration a magic lantern would be set up and "...As the shades of night begin to fall, the pictures... are shown on a screen filling up the end of the van". The slides for this show would have been produced by Watkins as were the 'Optical Lantern Readings' which accompanied them.

The Herefordshire Bee Van.

39

2nd Edition

OPTICAL LANTERN
READINGS.

⁂BEES⁂
AND
BEE·KEEPING,
By ALFRED WATKINS.

Hon. Sec. of the Herefordshire Bee-Keepers' Association.

ENTERED AT STATIONERS' HALL.

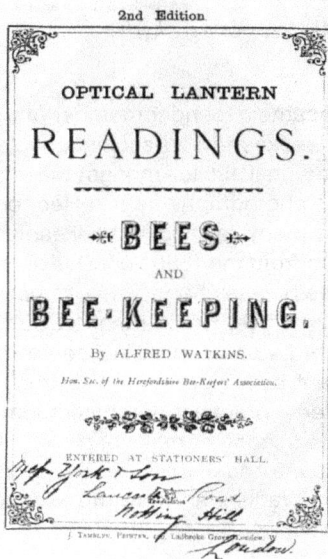

Watkin's "Optical Lantern Readings". As he was Hon. Sec. from 1882-1901, so presumably these notes were written during that period.

Watkins' presidential address to the Woolhope Naturalists Field Club in 1919 was on 'Bees and Beekeeping' about which he commented "the one natural-history subject on which I am sufficiently qualified to speak, that of the only insect which we in the British Isles subserve to the use of man—namely the Honey Bee." Having studied this address and his Lantern Readings, it is quite clear that Watkins had a great depth of knowledge of the subject matter and was well read. He was very familiar with the latest techniques and keen to promote moveable-frame hive beekeeping. He considered the 'old system' to be "...as a sealed book to the owner, and it was most difficult to provide extra room in the hive for the storage of honey and to prevent swarming".

One of Watkins' publications 'Must We Trade In Tenths' would probably have had support in a more recent era. He was vehemently opposed to a proposal to convert to a decimal currency and argued the case for 'octaval coinage'; a system based on the series 2,4, 8 etc., and the corresponding fractions of a half, quarter, eighth and so on. His idea gained support from no less a person than George Bernard Shaw!

One could go on and on listing the accomplishments of this fascinating character; his daughter observed how "...he wore (winter and summer) suits of Harris Tweed lined with grey flannel, containing fourteen pockets. These pockets were filled with letters, pamphlets, tools, rulers and other paraphernalia." The suits were as a metaphor of his life; in his younger days he was a medal-winning rower, he was considered a 'good amateur conjurer', an accomplished skater, he had a "considerable talent for wood-working", produced a prize-winning flour and invented a baker's thermometer. It was said of him that he was a "Highly **cultured man who knew** everything about something, and something about everything." For those readers who wish to know something more of this amazing man I recommend "ALFRED WATKINS A Herefordshire Man" from which I gleaned much of the material for this article. It was written by Ron Shoesmith and is obtainable from Amazon UK from £0.49

Alfred Watkins

A Herefordshire Man

Ron Shoesmith

*My thanks to Robin Hill of the Hereford and Worcester County Libraries **who supplied the** photographs.*

Ninemaidens

Mead

Award winning mead & honey

visit **www.ninemaidensmead.com**
or tel. 01209 820939 / 860630

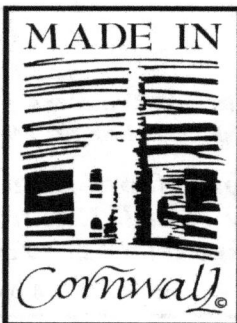

ALL IN A DAY'S WORK

BY DEREK LONEY - AN INTREPID
BEE BOLE HUNTER

I was on the trail of a 16th Century bee bole, but it looked as if my attempts to photograph and measure it would be thwarted. The gated entrance is at the foot of a short unmade road, to one side of which an unrelated farmhouse also faces. The heavy, double wrought iron gates are large, old but electronically controlled, and set within an 8ft boundary wall. A control button has an intercom unit above while in addition there is a small sign announcing that after 12.30pm the gates are locked so people should use the intercom to gain entry. I did, but got no reply. Twice. Couldn't quite believe that exit would not be possible during the afternoon for this didn't seem altogether usual, so I tried the release button – the gates slowly opened and I entered leaving my car parked out of the way outside. OK so far.

Down the curving driveway, past the stocks – no-one in them today, and there facing me was the facade of the C17 Hall in the near distance. On reaching it, whilst noting the security camera, I used the bell push at the large Victorian entrance porch which had been added to the main, older, structure. Tried it twice but with no response. So, after a few minutes and with a bit of

43

acting 'not sure quite what to do' for the camera, I made my way slowly back up the drive half wondering whether the door behind me might open after all. It didn't. Back at the gate I reached through to press the release button but to no effect. Tried it again; still no result.

I was effectively incarcerated. My first experience of a category C prison but with no means of help in evidence and little chance, it seemed, of remission or parole. Suddenly the afternoon seemed to have adopted a gloom or was it a doom I hadn't discerned earlier. The owner, whoever it might be, could be away for a month. I had noticed a post with a mobile telephone number a few yards along the drive – inside the gates - so I retraced my footsteps, noted the number and rang it from my mobile. No answer, although it did revert to an answerphone but that didn't seem to be of much help in my immediate predicament. I went back to the gates to find that there was a local landline number on the sign so I tried this. Each time it just clicked and went off!

Ah, I can't get a response from the house, from the mobile, nor from the unattributed landline number, so where do I go from here? Not, I decided, over the wall as it was high and I couldn't be sure what lay the other side.

Now then, there is high security, so perhaps the Police have a link or contact details, or might assist in some other way? Ladders could be a possibility but the Police helicopter would be of no use due to the heavy tree cover. Nevertheless I rang their number and explained to the police operator my problem. I also explained my purpose in having entered the grounds. "Well, why did you go in?" I was asked. Gently I explained again whilst acknowledging that they probably didn't receive such calls on a daily basis. "No. We don't." he said, "I'll check records and make a call or two to see what information we might have. Can you tell me the post code because we might need to send someone round?"

"Unfortunately, if I have that it'll be in my car which I left outside, this is the phone number I've tried but it doesn't connect."

"Will you hold on a few minutes then while I make a call and see whether I can locate the property where you are?"

"Yes of course", I agreed. "Thank you very much."

After a lengthy pause he came back to me saying that he had managed to make contact with a gentleman who would open the gates for me. "But he sounds very aggressive, so I'd suggest you try not to antagonise him further."

"Ah", I said, "Yes, he's just come to the door". Thanks ever so much."

"Oh, hullo, I'm so sorry to trou...."

"You shouldn't be here? This is private property! Get out!"

"Er, well, I was hoping to be able ....." but it was too late, with a: "I'm very busy!" he had already turned and was closing the door behind him, so slowly I retraced my steps to the gate which by that time had opened, exposing me to a welcoming world of freedom once more.

Outside at last, I pressed the intercom button, to hear an irrascible "What do you want, I told you this is private property!"

"Well, I just thought it would be a courtesy to let you know I had left, and to thank you for letting me out."

"Oh. Well you should make an appointment if you want to visit."

"Well, yes, but how does one do that? The phone number didn't connect when I tried it. I might not look it particularly today, but I do have my more respectable moments on some occasions. I had hoped I might be able to re-photograph your bee boles for the International Bee Research Association (IBRA)."

"You'd better come in. I'll meet you at the front door; I'm sorry I was short but I've been having a bad day."

The gates opened. I duly traipsed back up the drive to find a quite well-built man in his 60s looking at me enquiringly with one eye while keeping watch on all else with the other: I had a feeling it might be telescopic, probably with infra-red night vision as well.

I introduced myself and told him that I came from Hebden Bridge and explained my cause in more detail whereupon he apologised once more, telling me how he'd had a bit of an upset and it had been a bad day. He hadn't, he said, got to his mobile in time when I rang earlier because he was on the computer. He asked whether when I had been to the door earlier I had gone round to the garden, to which I said "Of course not". I simply hadn't realised I really wouldn't be able to depart! Much more relaxed by now he, for I had not enquired nor had he told me his name, explained I should go round the front of the house, across the old tennis court lawn from where I would see the greenhouse and the bee boles. I could photograph and record: if I would give a call at the door when I'd finished he would let me out. But, he said, he was too busy to accompany me. I showed him the 1979 photograph whereupon he remarked that I would find them little changed; he had redone the greenhouse recently but apart from that ... The walls round there, he told me, all date from the 16th century.

Round at the far side of the house I found the bee boles as described, photographed them and returned to the entrance having also established the 10-figure NGR and the altitude. He came out in response to my call. We then had quite a long and friendly discussion about the house and bee boles and he mentioned he had once visited Castle Carr, which he thought was near Hebden Bridge, as a member of a shooting party. "What a pity," he said, "that the house was demolished." "Ah yes, that was in the early 1960s." I said. He hadn't seen the Castle Carr fountains in action and asked if I had, which I was able to confirm and suggest that should he ever have the opportunity this is something he really shouldn't miss. Acerbic had become very affable.

He then mentioned the publication of a magazine article saying, "Stay there, and I'll show you." Three or four minutes later he returned to offer me a printout he had just made. This proved to be of an article in 'The Dalesman' of 1984 "Bee Boles in a Wharfedale Garden" which gave a short account of bee boles. Specifically, it mentions Ponden Hall which I visited two or three years ago, and holes at East Riddlestone Hall near Keighley which are now known to have housed falcons.

We chatted for some while before shaking hands and parting. I then repeated my trek along the drive with a glance of relief at the stocks before reaching the gates, only to find them shut fast and not responding to the operating button. Hesitantly, I pressed the intercom: nothing. I pressed again.

This time there was an answer: "Oh, sorry," he said, "I forgot."

With another word of thanks, and a sort of silent prayer, I went on my way – to undertake another visit another bee bole, mentioned in a 1957 magazine article.

(I have been acting as a recorder for the IBRA Bee Bole and Hive Shelter Register for the past few years and in that time have been privileged to be able to locate, assess and photograph a large number of sites in the north of England and Scotland, some of which have become 'lost' - right up to the north coast. The delights and satisfactions are multiple. Meeting many people who reflect how fortunate they are to be custodians of such structures, and others who are thrilled to learn what they have. Additionally there is sometimes the 'excitement of the chase' when the location is not certain plus, of course, the satisfaction of locating previously unrecorded sites).

Whilst IBRA hold a very good listing of bee boles and shelters and their locations, many of them are on private land and visitors are not necessarily welcome to view them, even if an attempt has been made to contact the owners before hand. Editor.

Bee Boles found in September include
An array at a bastle house near Catton in Northumberland; (b) a nice set of four at a cottage in the South Lakes (there is a further set of three on the other side of the wall); and (c) an array of six near the centre of Haltwhistle in Northumberland. I found the last two last August. All three are, of course, on private property.

OLD IS OFTEN BEST

John Kinross

●

The time has come again for the TARANOV Board to meet up in our local pub. For new readers it stands for "Tomes and Reprints avaiable, not on Varroa" and as Secretary I have to welcome Professor Drlpitoff who turns up on his bike, and Ina Strainer who comes in her ancient Morris Traveller, the back full of ekes, lambs wool, unused and cracked honey jars and her shopping from two days ago which she keeps meaning to take out of her car.

Everyone agrees on the new book of the year. This is Karl Showler's "Essays in Beekeeping History" (Bee Craft £17.50), 272 pages with some delightful line drawings. Ina and I think it should be at least £20 and that the picture of Mrs Hooper in her bridal gown (?), attempting to house a swarm, should have been on the cover. There are also pictures of Anna Comstock (in a long skirt and twin-strap shoes) reading, possibly her book, "How to keep Bees" in 1905 in USA. Ina enjoyed the chapter on women writers.The Professor enjoyed the chapter on Edward Bevan ("The Honey Bee" 1827) who lived locally and was a shareholder In the tram railway that linked Hay with the Abergavenny-Brecon canal and was driven by horsepower. This is a fascinating book that you can dip into and which helps prove that in beekeeping old is often best.

Aston and Bucknall have produced "Keeping Healthy Honey Bees" (NBB £16.00) following on from their book "Plants and Honeybees" (NBB 2009, £20) with the same sort of illustrations, small colour ones with longish blurbs. Emphasis is on the importance of healthy bees and there are useful points on inspectlng your hives, etc. One point they mention is that some hives are not easy to hold. The answer is to look at the Nutshell selectlon of booklets (edited by the author and produced for £1.50 each by NBB) and they wIll find a useful one on making a hlve carrier.

BBN0 have at long last sent new "Beekeeping Study Notes", Modules 1,2 & 3, to press (BBNO £30) and for those who, like the Professor, are concerned about Module 4, it has been discontinued by the Exam Board as few students bothered with it as most of the informatlon will be in the Orange book Modules 5-8 (BBNO £35.00). Ina likes the colours of the Study Notes and we argue about the Purple book (Husbandry Notes, BBNO £19.75) which she says is mauve.

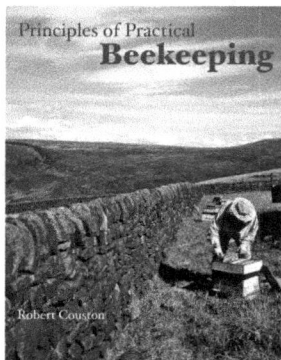

There is a Scottlsh element to two titles, one new, one a reprint. Bob Couston"s "Principles of Practical Beekeeping" (NBB £11.95) has long been out of print and the first reprint had some poor illustrations, as the book was orlglnally a series of newspaper articles. It came out in hardback in 1972 and first editions are now scarce, so all Scottish beginners need to get hold of a copy of this purple ('mauve', says Ina) -covered book. The other book is of great interest to all who know Chain Bridge Honey Farm at Berwick-on-Tweed. "Reflections on Beekeeping" (NBB £9.95) is Willie Robson's account of his experiences as a professional beekeeper. I recall first hearing him speak a few years ago. He said the Scottish

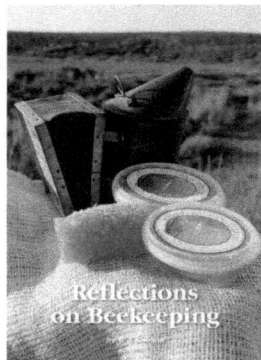

Inspectors called and he claimed all his hives were in Northumberland. When later the English inspectors called he claimed they were in Scotland. Berwick is part of England but Berwickshire is Scotland. Read the book and you won't be disappointed; there are even pictures of Willie's three tractors on the back but alas no picture of the double-decker bus that serves as a restarant on the south bank of the Chain Bridge.

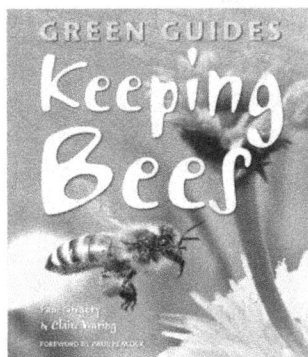

Finally a beginner's book which has only recently come our way. Pam Gregory, Claire Waring and Paul Peacock have joined up in writing "Green Guides - Keeping Bees". It is a 256pp paperback which is excellent value for £9.99.The lay-out is excellent and I like the "Hot Tips".The Professor especially likes the one on p.160 "Make sure the escape is the right way up" as it reminds him of his Harley motor cycle days. His friend Bob fell off one day and a passer-by in a car stopped to see if he was all right. When the Professor found Bob he asked the car driver if he was all right. "He was", said the man,"until we tried to put his helmet on the right way round."

WORD SEARCH

There are twenty-five species/races of honeybees hidden in the grid below, which can be found by following lines in any straight direction.

1.	14.
2.	15.
3.	16.
4.	17.
5.	18.
6.	19.
7.	20.
8.	21.
9.	22.
10.	23.
11.	24.
12.	25.
13.	

SPECIES OF HONEYBEES

```
A  A  A  F  W  A  M  R  T  A  I  S  M  A  S
K  C  A  D  L  K  O  E  C  T  A  T  O  T  I
A  E  I  U  A  L  S  A  D  H  T  C  N  A  S
F  C  C  N  O  N  I  U  A  A  A  I  T  L  N
E  I  I  C  R  L  S  R  U  U  I  B  I  L  E
S  A  I  T  O  A  A  O  C  R  R  E  C  E  P
O  N  P  T  S  I  C  A  N  X  A  R  O  T  A
U  K  A  O  E  U  S  I  E  I  N  I  L  U  C
R  N  Y  N  S  I  G  S  T  B  I  C  A  C  A
A  M  S  N  C  K  Z  I  Y  I  V  A  Z  S  I
K  I  T  A  D  A  M  I  L  R  N  J  G  M  R
S  L  A  M  A  R  C  K  I  I  I  E  E  I  P
A  I  S  S  I  M  R  E  T  N  I  A  M  V  Y
A  R  E  F  I  L  L  E  M  G  M  Y  C  E  C
A  I  W  G  V  S  L  I  T  O  R  E  A  A  J
```

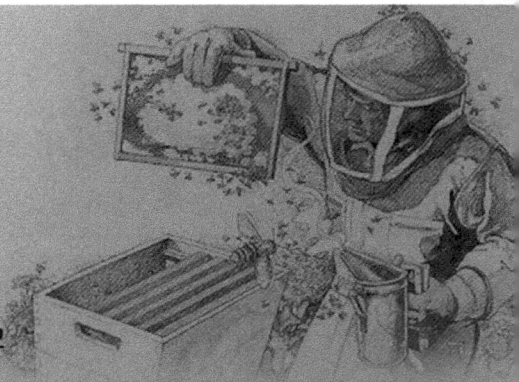

DIARY & CALENDAR

13

- PART II -

*SR (SUNRISE) SS (SUNSET) FOR LONDON UK.

JANUARY

That which is not good for the bee-hive cannot be good for the bees."
Marcus Aurelius (AD 121 -180), Philosopher and Emperor,

DAY	JANUARY 2013 FORAGE	TEMP MIN	MAX	WIND DIR	B.S	CL'D	RAIN	1	2	3 HIVE WEIGHT
1										
2										
3										
4										
5										
6										
7										
8										
9										
10										
11										
12										
13										
14										
15										
16										
17										
18										
19										
20										
21										
22										
23										
24										
25										
26										
27										
28										
29										
30										
31										

JAN13

	8,TU
1,TU NEW YEAR'S DAY	**9,WE**
2,WE BANK HOLIDAY	**10,TH**
3,TH	**11,FR** ○
4,FR	**12,SA** SR:08:02, SS:16:17
5,SA SR 05:23, SS 20:32	**13,SU**
6,SU	**14,MO**
7,MO	**15,TU**

16,WE	**24,TH**
17,TH	**25,FR**
18,FR	26,SA SR:07:47, SS:16.39
19,SA SR:07:56, SS:16:28	27,SU ●
20,SU	**28,MO**
21,MO	**29, TU**
22,TU	**30, WE**
23,WE	**31, TH**

FEBRUARY

Canterbury:
"...for so work the honey-bees,
Creatures that by a rule in nature teach
The act of order to a peopled kingdom.
They have a king and officers of sorts;
Where some, like magistrates, correct at home,
Others, like merchants, venture trade abroad,
Others, like soldiers, armed in their stings,
Make boot upon the summer's velvet buds,
Which pillage they with merry march bring home
To the tent-royal of their emperor;
Who, busied in his majesty, surveys
The singing masons building roofs of gold,
The civil citizens kneading up the honey,
The poor mechanic porters crowding in
Their heavy burdens at his narrow gate,
The sad-eyed justice, with his surly hum,
Delivering o'er to executors pale
The lazy yawning drone."

William Shakespeare (26th April 1564 (baptised) – 23rd April 1616),
from "Henry the Vth"

DAY	FEBRUARY 2013 FORAGE	TEMP		WIND		CL'D	RAIN	1	2	3
		MIN	MAX	DIR	B.S			HIVE WEIGHT		
1										
2										
3										
4										
5										
6										
7										
8										
9										
10										
11										
12										
13										
14										
15										
16										
17										
18										
19										
20										
21										
22										
23										
24										
25										
26										
27										
28										

FEB13

	8,FR
1,FR	**9,SA** SR:07:25, SS:17:05
2,SA CANDLEMAS DAY SR:07:37, SS:16:52	**10,SU** ○ CHINESE NEW YEAR - THE YEAR OF THE SNAKE
3,SU	**11,MO**
4,MO ●	**12,TU**
5,TU	**13,WE**
6,WE	**14,TH** ST VALENTINE'S DAY
7,TH	**15,FR**

16,SA SR:07:12, SS:17:18	24,SU *
17,SU	25,MO ●
18,MO	26,TU
19,TU ○	27,WE
20,WE	28,TH
21,TH	
22,FR	
23,SA SR:06:58, SS:17:30	

MARCH

"Even bees, the little almsmen of spring bowers,
know there is richest juice in poison-flowers."

John Keats (31st October 1795 – 23rd February 1821), from "Isabella".

DAY	MARCH 2013 FORAGE	TEMP		WIND		CL'D	RAIN	1	2	3
		MIN	MAX	DIR	B.S			HIVE WEIGHT		
1										
2										
3										
4										
5										
6										
7										
8										
9										
10										
11										
12										
13										
14										
15										
16										
17										
18										
19										
20										
21										
22										
23										
24										
25										
26										
27										
28										
29										
30										
31										

MAR13

	8,FR
1,FR ST DAVID'S DAY	9,SA SR:06:28, SS:17:55
2,SA SR:06:43, SS:17:43	10,SU
3,SU	**11,MO** ○
4,MO ●	**12,TU**
5,TU	**13,WE**
6,WE	**14,TH**
7,TH	**15,FR**

16,SA SR:06:12, SS:18:07	24,SU *
17,SU ST PATRICK'S DAY - BANK HOLIDAY (NORTHERN IRELAND)	25,MO
18,MO	26,TU
19,TU	27,WE ●
20,WE	28,TH
21,TH	29,FR GOOD FRIDAY
22,FR	30,SA SR:05:40, SS:18:31
23,SA SR:05:56, SS:18:19	31,SU EASTER DAY

APRIL

"A flock of sheep that leisurely passes by
One after one: the sound of rain, and bees
Murmuring: the fall of rivers, winds and seas.
Smooth fields, white sheets of water, and pure sky:
I have thought of all in turn, and yet do lie Sleepless!"

William Wordsworth (7th April 1770 - 23rd April 1850), from "To Sleep".

DAY	APRIL 2013 FORAGE	TEMP MIN	MAX	WIND DIR	B.S	CL'D	RAIN	1	2 HIVE WEIGHT	3
1										
2										
3										
4										
5										
6										
7										
8										
9										
10										
11										
12										
13										
14										
15										
16										
17										
18										
19										
20										
21										
22										
23										
24										
25										
26										
27										
28										
29										
30										

APR13

	8,MO
1,MO BANK HOLIDAY EASTER MONDAY	**9,TU**
2,TU	**10,WE** ○
3,WE	**11,TH**
4,TH	**12,FR**
5,FR	13,SA SR:06:09, SS:19:54
6,SA SR:06:24, SS:19:42	14,SU *
7,SU	**15,MO**

16,TU	**24,WE**
17,WE	**25,TH** ●
18,TH	**26,FR**
19,FR *	27,SA SR:05:40, SS:20:17
20,SA * SR:05:54, SS:20:06	28,SU
21,SU *	**29,MO**
22,MO	**30,TU**
23,TU	* BBKA SPRING CONVENTION, HARPER ADAMS UNIVERSITY COLLEGE CAMPUS, NEWPORT, SHROPSHIRE TF10 8NB

"The air came laden with the fragrance it caught upon its way, and the bees, upborne upon its scented breath, hummed forth their drowsy satisfaction as they floated by."

Charles Dickens (7th February 1812 - 9th June 1870)
from "The Old Curiosity Shop".

DAY	MAY 2013 FORAGE	TEMP		WIND		CL'D	RAIN	1	2	3
		MIN	MAX	DIR	B.S			HIVE WEIGHT		
1										
2										
3										
4										
5										
6										
7										
8										
9										
10										
11										
12										
13										
14										
15										
16										
17										
18										
19										
20										
21										
22										
23										
24										
25										
26										
27										
28										
29										
30										
31										

MAY13

	8,WE
1,WE	**9,TH**
2,TH ●	**10,FR** ○
3,FR	11,SA SR:05:15, SS:20:40
4,SA SR:05:27, SS:20:29	12,SU
5,SU	**13,MO**
6,MO EARLY MAY BANK HOLIDAY	**14,TU**
7,TU	**15,WE**

16,TH	**24,FR**
17,FR	**25,SA** ● SR:04:56, SS:21:00
18,SA SR:05:04, SS:20:50	26,SU
19,SU	**27,MO** SPRING BANK HOLIDAY
20,MO	**28,TU**
21,TU	**29,WE**
22,WE	**30,TH**
23,TH	**31,FR** ●

JUNE

"To make a prairie it takes a clover and one bee,—
One clover, and a bee,
And revery.
The revery alone will do
If bees are few."

Emily Dickinson (10th December 1830 - 15th May 1886), from "Nature No 97".

DAY	JUNE 2013 FORAGE	TEMP MIN	TEMP MAX	WIND DIR	WIND B.S	CL'D	RAIN	1	2	3
								HIVE WEIGHT		
1										
2										
3										
4										
5										
6										
7										
8										
9										
10										
11										
12										
13										
14										
15										
16										
17										
18										
19										
20										
21										
22										
23										
24										
25										
26										
27										
28										
29										
30										

JUN13

	8,SA ○ SR:04:45, SS:21:15
1,SA SR:04:49, SS:21:08	**9,SU**
2,SU	**10,MO**
3,MO	**11,TU**
4,TU	**12,WE**
5,WE	**13,TH** ○
6,TH	**14,FR**
7,FR	**15,SA** SR:04:43, SS:21:19

16,SU	24,MO
17,MO	25,TU
18,TU	26,WE
19,WE	27,TH
20,TH	28,FR
21,FR SUMMER SOLSTICE	29,SA SR:04:46, SS:21:21
22,SA SR:04:43, SS:21:22	30,SU
23,SU ●	

JULY

"He said the pleasantest manner of spending a hot July day was lying from morning till evening on a bank of heath in the middle of the moors, with the bees humming dreamily about among the bloom, and the larks singing high up overhead, and the blue sky and bright sun shining steadily and cloudlessly."

Emily Bronte (30th July 1818 - 19th December 1848), from "Wuthering Heights".

Portrait: - Anne, Emily and Charlotte, painted by their brother Branwell.

DAY	JULY 2013 FORAGE	TEMP		WIND		CL'D	RAIN	1	2	3
		MIN	MAX	DIR	B.S			HIVE WEIGHT		
1										
2										
3										
4										
5										
6										
7										
8										
9										
10										
11										
12										
13										
14										
15										
16										
17										
18										
19										
20										
21										
22										
23										
24										
25										
26										
27										
28										
29										
30										
31										

JUL13

	8,MO ○
1,MO	**9,TU** RAMADAN BEGINS
2,TU	**10,WE**
3,WE	**11,TH**
4,TH	**12,FR**
5,FR	13,SA SR:04:59, SS:21:16
6,SA SR:04:52, SS:21:18	14,SU ST SWITHUN'S DAY
7,SU	**15,MO** ○

16,TU	**24,WE** FIBKA - GORMANSTON CONVENTION
17,WE	**25,TH** FIBKA - GORMANSTON CONVENTION
18,TH	**26,FR** FIBKA - GORMANSTON CONVENTION
19,FR	27,SA SR:05:17, SS:20:56 FIBKA - GORMANSTON CONVENTION
20,SA SR:05:07, SS:21:06	28,SU FIBKA - GORMANSTON CONVENTION
21,SU	**29,MO**
22,MO ●	**30,TU**
23,TU	**31,WE**

AUGUST

"There's a whisper down the field where the year has shot her yield,
And the ricks stand grey to the sun,
Singing: 'Over then, come over, for the bee has quit the clover,
'And your English summer's done.' "

Rudyard Kipling (30th December 1865 - 18th January 1936) from "The Long Trail"

DAY	AUGUST 2013 FORAGE	TEMP		WIND		CL'D	RAIN	1	2	3
		MIN	MAX	DIR	B.S			HIVE WEIGHT		
1										
2										
3										
4										
5										
6										
7										
8										
9										
10										
11										
12										
13										
14										
15										
16										
17										
18										
19										
20										
21										
22										
23										
24										
25										
26										
27										
28										
29										
30										
31										

AUG13

	8,TH
1,TH	**9,FR**
2,FR	**10,SA** SR:05:38, SS:20:32
3,SA SR:05:27, SS:20:45	**11,SU**
4,SU	**12,MO** SR 05:42, SS 20:28 FIRST DAY OF GROUSE SHOOTING - TRADITIONAL TIME FOR TAKING BEES TO THE HEATHER MOORS.
5,MO	**13,TU**
6,TU ○	**14,WE**
7,WE RAMADAN FINISHES	**15,TH**

16,FR	24,SA ST BARTHOLOMEW'S DAY, TRADITIONAL DAY FOR HARVESTING HONEY SR:06:00, SS:20:04
17,SA SR:05:49, SS:20:19	25,SU
18,SU	**26,MO** SUMMER BANK HOLIDAY
19,MO	**27,TU**
20,TU	**28,WE**
21,WE ●	**29,TH**
22,TH	**30,FR**
23,FR	31,SA SR:06:12, SS:19:49

SEPTEMBER

"Stands the Church clock at ten to three?
And is there honey still for tea?"

*Rupert Brooke (3rd August 1887 - 23rd April 1915)
from the "The Old Vicarage, Grantchester," 1912.*

DAY	SEPTEMBER 2013 FORAGE	TEMP MIN	MAX	WIND DIR	B.S	CL'D	RAIN	1	2	3 HIVE WEIGHT
1										
2										
3										
4										
5										
6										
7										
8										
9										
10										
11										
12										
13										
14										
15										
16										
17										
18										
19										
20										
21										
22										
23										
24										
25										
26										
27										
28										
29										
30										

SEP13

	8,SU
1,SU	**9,MO**
2,MO	**10,TU**
3,TU ●	**11,WE**
4,WE	**12,TH**
5,TH ○	**13,FR**
6,FR	**14,SA** SR:06:34, SS:19:17
7,SA SR:06:23, SS:19:33	**15,SU** 33RD APIMONDIA CONGRESS, KIEV, UKRAINE

16,MO	**24,TU**
17,TU	**25,WE**
18,WE	**26,TH**
19,TH ●	**27,FR**
20,FR	28,SA SR:06:56, SS:18:45 33RD APIMONDIA CONGRESS, KIEV, UKRAINE
21,SA SR:06:45, SS:19:01	29,SU 33RD APIMONDIA CONGRESS, KIEV, UKRAINE
22,SU	**30,MO** 33RD APIMONDIA CONGRESS, KIEV, UKRAINE
23,MO AUTUMN EQUINOX	

OCTOBER

"When I hear a man preach, I like to see him act as if he were fighting bees."

Abraham Lincoln (12th February 1809 - 15th April 1865).

DAY	OCTOBER 2013 FORAGE	TEMP MIN	MAX	WIND DIR	B.S	CL'D	RAIN	1	2	3 HIVE WEIGHT
1										
2										
3										
4										
5										
6										
7										
8										
9										
10										
11										
12										
13										
14										
15										
16										
17										
18										
19										
20										
21										
22										
23										
24										
25										
26										
27										
28										
29										
30										
31										

OCT13

	8,TU
1,TU 33RD APIMONDIA CONGRESS, KIEV, UKRAINE	**9,WE**
2,WE 33RD APIMONDIA CONGRESS, KIEV, UKRAINE	**10,TH**
3,TH 33RD APIMONDIA CONGRESS, KIEV, UKRAINE	**11,FR**
4,FR 33RD APIMONDIA CONGRESS, KIEV, UKRAINE	**12,SA** SR:07:19, SS:18:13
5,SA ○ SR:07:08, SS:18:29	13,SU
6,SU	**14,MO**
7,MO	**15,TU**

16,WE	**24,TH** NATIONAL HONEY SHOW
17,TH	**25,FR** NATIONAL HONEY SHOW
18,FR ●	26,SA NATIONAL HONEY SHOW SR:07:44, SS:17:44
19,SA SR:07:31, SS:17:59	27,SU CLOCKS GO BACK ONE HOUR
20,SU	**28,MO**
21,MO	**29, TU**
22,TU	**30, WE**
23,WE	**31, TH**

NOVEMBER

"No shade, no shine, no butterflies, no bees
No fruits, no flowers, no leaves, no birds.
November."

Thomas Hood (17th May 1799 - 3rd May 1845) from "No!".

DAY	NOVEMBER 2013 FORAGE	TEMP MIN	MAX	WIND DIR	B.S	CL'D	RAIN	1	2 HIVE WEIGHT	3
1										
2										
3										
4										
5										
6										
7										
8										
9										
10										
11										
12										
13										
14										
15										
16										
17										
18										
19										
20										
21										
22										
23										
24										
25										
26										
27										
28										
29										
30										

NOV13

	8,FR
1,FR	9,SA SR:07:08, SS:16:20
2,SA SR:06:56, SS:16:31	10,SU
3,SU ○	**11,MO**
4,MO	**12,TU**
5,TU	**13,WE**
6,WE	**14,TH**
7,TH	**15,FR**

16,SA SR:07:20, SS:16:10	24,SU *
17,SU ●	25,MO
18,MO	26,TU
19,TU	27,WE
20,WE	28,TH
21,TH	29,FR
22,FR	30,SA SR:07:42, SS:15:56
23,SA SR:07:32, SS:16:01	

DECEMBER

"Like the bee, we should make our industry our amusement."

Oliver Goldsmith (10th November 1730 - 4th April 1774).
Statue of Goldsmith is in front of Trinity College, Dublin.

DAY	DECEMBER 2013 FORAGE	TEMP MIN	MAX	WIND DIR	B.S	CL'D	RAIN	1	2	3 HIVE WEIGHT
1										
2										
3										
4										
5										
6										
7										
8										
9										
10										
11										
12										
13										
14										
15										
16										
17										
18										
19										
20										
21										
22										
23										
24										
25										
26										
27										
28										
29										
30										
31										

DEC13

	8,SU
1,SU	9,MO
2,MO	10,TU
3,TU ○	11,WE
4,WE	12,TH
5,TH	13,FR
6,FR	14,SA SR:07:59, SS:15:51
7,SA ST AMBROSE DAY - PATRON SAINT OF BEEKEEPERS SR:07:52, SS:15:52	15,SU

16,MO	**24,TU**
17,TU ●	**25,WE** CHRISTMAS DAY BANK HOLIDAY
18,WE	**26,TH** BOXING DAY BANK HOLIDAY
19,TH	**27,FR**
20,FR	28,SA SR:08:06, SS:15:58
21,SA WINTER SOLSTICE SR:08:04, SS:15:53	29,SU
22,SU	**30,MO**
23,MO	**31,TU**

Hive/ Q NO.	Year Q Raised	Frames of Brood Autumn 2012	Combs Covered	Honey Stored- Sugar fed Kg	Combs Covered Spring 2013	Frames of Brood Spring 2013	Spring Feeding Kg	Queens Reared	Nuclei
1									
2									
3									
4									
5									
6									
7									
8									
9									
10									
11									
12									
13									
14									
15									
16									
17									
18									
19									
20									
21									
22									
23									
24									

HONEYBEE COLONIES

1									
2									
3									
4									
5									
6									
7									
8									
9									
10									
11									
12									
13									
14									
15									
16									
17									
18									
19									
20									
21									
22									
23									
24									

BEEEKEEPING RECORDS

Number	items	£	P
		Est. Value	
	Stocks of Bees		
	Empty Hives		
	Combs - Deep - Shallow		
	Frames		
	Foundations		
	Honey Extractor		
	Honey Tanks		
	Other items		
	Honey Jars		
	Honey		

JANUARY 2014						
S	M	T	W	T	F	S
			1	2	3	4
5	6	7	8	9	10	11
12	13	14	15	16	17	18
19	20	21	22	23	24	25
26	27	28	29	30	31	

FEBRUARY 2014						
S	M	T	W	T	F	S
						1
2	3	4	5	6	7	8
9	10	11	12	13	14	15
16	17	18	19	20	21	22
23	24	25	26	27	28	

MARCH 2014						
S	M	T	W	T	F	S
						1
2	3	4	5	6	7	8
9	10	11	12	13	14	15
16	17	18	19	20	21	22
23	24	25	26	27	28	29
30	31					

APRIL 2014						
S	M	T	W	T	F	S
		1	2	3	4	5
6	7	8	9	10	11	12
13	14	15	16	17	18	19
20	21	22	23	24	25	26
27	28	29	30			

MAY 2014						
S	M	T	W	T	F	S
				1	2	3
4	5	6	7	8	9	10
11	12	13	14	15	16	17
18	19	20	21	22	23	24
25	26	27	28	29	30	31

JUNE 2014						
S	M	T	W	T	F	S
1	2	3	4	5	6	7
8	9	10	11	12	13	14
15	16	17	18	19	20	21
22	23	24	25	26	27	28
29	30					

JULY 2014						
S	M	T	W	T	F	S
		1	2	3	4	5
6	7	8	9	10	11	12
13	14	15	16	17	18	19
20	21	22	23	24	25	26
27	28	29	30	31		

AUGUST 2014						
S	M	T	W	T	F	S
					1	2
3	4	5	6	7	8	9
10	11	12	13	14	15	16
17	18	19	20	21	22	23
24	25	26	27	28	29	30
31						

SEPTEMBER 2014						
S	M	T	W	T	F	S
	1	2	3	4	5	6
7	8	9	10	11	12	13
14	15	16	17	18	19	20
21	22	23	24	25	26	27
28	29	30				

OCTOBER 2014						
S	M	T	W	T	F	S
			1	2	3	4
5	6	7	8	9	10	11
12	13	14	15	16	17	18
19	20	21	22	23	24	25
26	27	28	29	30	31	

NOVEMBER 2014						
S	M	T	W	T	F	S
						1
2	3	4	5	6	7	8
9	10	11	12	13	14	15
16	17	18	19	20	21	22
23	24	25	26	27	28	29
30						

DECEMBER 2014						
S	M	T	W	T	F	S
	1	2	3	4	5	6
7	8	9	10	11	12	13
14	15	16	17	18	19	20
21	22	23	24	25	26	27
28	29	30	31			

All efforts have been made to ensure the accuracy of the information in these pages. Corrections and amendments should be sent to The Editor The Beekeepers Annual, c/o Northern Bee Books, Scout Bottom Farm, Mytholmroyd, Hebden Bridge HX7 5JS

DIRECTORY, ASSOCIATIONS AND SERVICES

DIRECTORY, ASSOCIATIONS AND SERVICES

BEE MAILING

BEEKEEPING MAILING LISTS

http://www.zbee.dircon.co.uk

Beekeeping mailing list services provided by zbee.com http://www.zbee. dircon.co.uk

KENT BEEKEEPERS ASSOCIATION, THE
Name of mailing list: Kentbee-L Serving a possible membership of 400. **Support website:** http://www.kentbee.com Approximately 80 have subscribed. Providing a forum for local branch announcements and news and chat about beekeeping. **To subscribe to Kentbee-L send a message to:** mailserver@zbee.com **Subject field:** You leave this blank it doesn't matter. **In the message body write:** Subscribe Kentbee-L then send the message and await further instructions to complete the subscription process.

NATIONAL HONEY SHOW, THE
Name of mailing list: NHS The National Honey Show is held in October each year in London, the support website http://www.honeyshow.co.uk has more information and schedules, **To subscribe to NHS send a message to:** mailserver@zbee.com, **Subject field:** You leave this blank it doesn't matter., **In the message body write:** Subscribe NHS then send the message and await further instructions to complete the subscription process

BEE IMPROVEMENT & BEE BREEDERS ASSOCIATION, THE (BIBBA)
Name of mailing list: BIBBA-L, Support website http://www. bibba.com/, **To subscribe to BIBBA-L send a message to:** mailserver@zbee.com, **Subject field:** You leave this blank it doesn't matter. **In the message body write:** Subscribe BIBBA-L then send the message and await further instructions to complete the subscription process.

BEE MAILING

✉ ☎

APINET (BEEKEEPING EDUCATION EXTENSION NETWORK)
Name of mailing list: APINETL, Support website n/a, **To subscribe send a message to:** mailserver@zbee.com, **Subject field:** You leave this blank it doesn't matter.
In the message body write: Subscribe APINETL then send the message and await further instructions to complete the subscription process.

BROMLEY & SIDCUP & ORPINGTON BEEKEEPERS ASSOCIATION
Name of mailing list: BBK, **Support website:** http://www.kentbee.com/, **To subscribe to BBK send a message to:** mailserver@zbee.com, **Subject field:** You leave this blank it doesn't matter. **In the message body write:** Subscribe BBK then send the message and await further instructions to complete the subscription process.

THE BRITISH BEEKEEPERS ASOCIATION (BBKA)
Name of mailing list: BBKA, **Support website:** http://www.bbka.org.uk, Private list members only, see members area for joining details.

BDI

BDI

BEE DISEASES INSURANCE LTD

SECRETARY
Donald Robertson-Adams
Bryngwrog
Beulah, Newcastle Emlyn
Ceredigion, SA38 9QR
07532 336076
donald@theoldmill.fsnet.
co.uk

Bee Diseases Insurance (BDI) provides insurance cover for individual beekeepers, association apiaries and commercial beekeepers alike, against the possibility of their bees and equipment being destroyed as a result of a Destruction Order following a visit from an authorised Bee Inspector. .

BDI provides compensation for specified property that may need to be destroyed as a result of American Foul Brood and European Foul Brood.

TREASURER AND
SCHEME B MANAGER
Mrs Sharon Blake
Stratton Court,
South Petherton,
Somerset TA13 5LQ
01460 242124
sbeditor@yahoo.co.uk

BDI has established a contingency fund capped at £25,000 a year if Small Hive Beetle or Tropilaelaps infestation is found.

Scheme A provides cover for the beekeeper with a total of 39 colonies or less. Cover is obtained by being a member of a Beekeeping Association that is a member of BDI Ltd.

Scheme B provides cover for beekeepers with 40 or more colonies in total. Insurance under this Scheme is on a personal basis and further details can be obtained from the Scheme B Manager.

CLAIMS MANAGER
Bernard Diaper
57 Marfield Close,
Walmley,
Sutton Coldfield B76 1YD
0121 3133112
b.diaper@tiscali.co.uk

REMEMBER: DISEASE CAN STRIKE ANY COLONY AT ANY TIME AND IT IS SPREAD THROUGHOUT THE COUNTRY. PROTECT YOUR APIARY, AND OTHER BEEKEEPERS, THROUGH B.D.I.

PRESIDENT
Richard Ball
Stoneyford Farmhouse
Colaton Raleigh
Sidmouth
Devon, EX10 0HZ
01395 567990
richard.ball@fera.gsi.
gov.uk

BEE FARMERS' ASSOCIATION OF THE UNITED KINGDOM

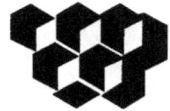

The BFA represents the professional beekeepers of the UK.

The association is the largest contract pollinator in the UK and our members are responsible for virtually all the migratory pollination. They are expected to have a good degree of competence; membership requires over 40 hives, and sponsorship by a BFA member who knows the applicant as a beekeeper. We have recently introduced a code of conduct which members are expected to observe. In addition we have a significant number of members who get some income from being bee inspectors, responsible for identifying and dealing with notifiable disease.

Business is conducted at twice-yearly regional meetings which pass items up to the main meeting for discussion and voting, and which put forward candidates for the committee.

The BFA is affiliated to the National Farmers Union and The Honey (Packers) Association with whom we work effectively in promoting ecological sensitive farming and in promoting consumer awareness through events such 'National Honey Week' and bulk sales to retail chains.

MEMBERSHIP

Our members are expected to have a good degree of competence.

FULL MEMBERSHIP requires over 40 hives, and sponsorship by a BFA member who knows the applicant as a beekeeper.

ASSOCIATE MEMBERSHIP is a stepping stone to full membership of the BFA for beekeepers with a minimum of 20 hives and who would like to take up commercial or semi-commercial beekeeping.

Membership forms are available from the Membership Secretary, or as a download from our website.

FUNCTIONS

* To monitor and to keep members informed about developments in commercial beekeeping, bee science and UK and EEC legislation.
* Liaison with Farmers, Growers, Contractors, Consumers and other organisations.

CHAIRMAN
Murray McGregor
Denrosa Apiaries
Victoria Street
Coupar Angus
PERTHSHIRE
PH13 9AE
01828 627721
murray@denrosa.demon.co.uk

TREASURER,
Doug Isles
Hudnalls Apiary
Balligan Cottage
The Hudnalls
St. Briavels
Lydney
GLOUCESTERSHIRE
GL15 6RT
01594 530807
info@hudnallsapiaries.co.uk

GENERAL SECRETARY,
Margaret Ginman
Hendal House
Hendal Hill
Groombridge
Tunbridge Wells
KENT
TN3 9NT
01892 864 499 /
07795 153 765
margaret.hendal@btconnect.com

BFA

✉ ☎

POLLINATION SECRETARY
Alan Hart
61 Fakenham Road,
Great Witchingham,
Norwich,
NORFOLK
NR9 5AE
01603 308911
earlswoodbees@hotmail.
co.uk

RESEARCH AND
ADMINISTRATION
(INCLUDING MEMBERSHIP,
INSURANCE AND BULLETIN)
David Bancalari
Park Farm Barn
Shortthorn Road
Stratton Strawless
NORFOLK
NR10 5NX
01603 755105
wiredbrain@btinternet.com

- Liaison with UK Government Departments dealing with beekeeping, medicines, and allied matters.
- Liaison and co-operation with UK Beekeeping organisations.
- Contact with European beekeeping organisations (EPBA) and representation on the EEC Honey Working Party (COPA/ COGECA) in Brussels.
- Political lobbying through MPs and Euro MPs.
- Member of the Confederation of National Beekeeping Associations (CONBA)
- Member of the European Professional Beekeepers' Association (EPBA)
- Associate member of the Honey Association

FACILITIES FOR MEMBERS:

- Bi-monthly Bulletins with news and updates, notes on meetings with DEFRA, Fera, VMD, and the EEC, reports on current beekeeping problems (e.g. varroa) and commercial developments world-wide.. This bulletin is available as a paper and/or an e-document
  * e-news. Frequent electronic updates on news items
- Free advertisement of members' sales and wants (including hive products, bee stocks and spare equipment).
* Regional meetings which provide for local discussion and opportunities for trading between members.
- Crop and winter loss reports.
- Free Circulation among members of UK and foreign magazines.
- Free insurance for products and third party liability (not limited to thirty hives).
- Special rates for employers liability insurance.
- Comprehensive special beefarmers insurance with the NFU.
- Pollination contracts.
- Advice from experienced members on all aspects of honey farming and commercial beekeeping; sources of equipment and sundries.
- Product directory listing specialist suppliers.
- Discounts from suppliers.
- Bulk purchase schemes to minimize costs to individual members..

ANNUAL CONVENTION WEEKEND

- Spring meeting for members and partners, held each March at different locations in the UK or abroad. Visits to local bee and research establishments; lectures and discussions on bee-related matters; sight seeing, and social events.

BCS

BEEKEEPING COURSES & SERVICES

PART TIME LECTURERS & FURTHER EDUCATION COURSES IN BEEKEEPING

The following may offer a range of theoretical and practical courses in beekeeping, in some cases an advisory service or a diagnostic service for adult bee diseases only may be offered.

The range of services and activities is wide and this list is not exhaustive but the following may be contacted for details of facilities in an enquirer's area.

BEDFORDSHIRE,
Mike Nieman
43 Flitwick Road
Westoning
Bedfordshire
MK45 5JA
01525 717040
Harry Inman
10 Constable Hill
Bedford
MK41 7LJ
01234 306554
* Beekeeping for Beginners
* Practical Beekeeping

BERKSHIRE COLLEGE OF AGRICULTURE, Kate Malenczuk and Reg Hook
Hall Place,
Burchetts Green
Maidenhead,
Berkshire SL6 6QR
www.bca.ac.uk

Tel: 01628 824444
Fax 01628 827488
• Beekeeping for beginners
• Practical Beekeeping
• Preparing Bees for Winter
• Intermediate Beekeeping
• Taster Sessions
• A Top up for beginners at the end of their first season
• Further techniques for more experienced beekeepers.

CHESHIRE BEEKEEPERS (STOCKPORT BRANCH)
• Introduction To Beekeeping Course,
• Practical Beekeeping,
Stockport :-
Mrs Carolin Hallworth
01625 875 436
North Cheshire (Frodsham):-
Dan Fox 01565 777 341

South Cheshire (Bradwall):-
Mrs Liz Camm 01270 664 337
Wirral
Doug Jones (Thornton Hough)
0151 342 7062

DEVON, Dr. Mick Street
c/o Bicton College
Budleigh Salterton
EX9 7BY
or the DBKA Education
Officer at;
www.devonbeekeepers.
co.uk

ESSEX, Richard Ridler
Treasurer,
C/O Saffron Walden Division,
Essex Beekeepers'
Association,
Rundle House,
High Street
Hatfield Brand Oak,
Bishop's Stortford
Hertfordshire
CM22 7HE
richard.ridler@uwclub.net
01279 718111
07942 815753

LEEDS
For Details See YBKA

115

BCS

✉ ☎

NORFOLK, Paul Metcalf NDB
Easton College, Easton
Norwich NR9 8DX
SUSSEX, Business Training
Plumpton College
Ditchling Road
Plumpton, Nr Lewes
East Sussex BN7 3AE
pd@plumpton.ac.uk

STOCKPORT (see Cheshire)

WILTSHIRE,The Secretary
Melksham Beekeepers
Association
Deans End
Butts Lane
Keevil,
Trowbridge
BA14 6LZ
wickhamsoftkeevil@
btinternet.com

YBKA, Bill Cadmore,
104 Hall Lane, Horsforth
Leeds LS18 5JG
0113 216 0482
bill.cadmore@ntlworld.com
Venue:
Home Farm Rare Breeds Centre
Temple Newsam House
Garden & Estate East Leeds

116

BEEKEEPING EDITORS' EXCHANGE SCHEME

BEES is a self-help grouping of local, county and country beekeeping association editors, which operates principally by exchanging journals through a central address. The scheme is supported by Northern Bee Books.

BEES was founded in 1984 and for many years has been an exchange of paper copy. However, the focus has now changed to an electronic exchange, using the server of one of the participating editors.

Now fully established as part of the British and Irish beekeeping scene, the scheme brings up to date information to beekeepers throughout the British Isles.

B.E.E.S
Helping Editors
Help Themselves

Sponsored by
NORTHERN BEE BOOKS

The aims are:
- to exchange ideas for content and production methods
- to aid others by experience
- to communicate matters editorial
- to share information on national beekeeping issues
- to help and reassure those new to the task
- to give a wider readership to the best writing in beekeeping journalism

If you are an editor or potential editor and would like to know more about how we operate write to Chris Jackson
22 Chapter Close, Oakwood, Derby, DE21 2BG
editors-owner@ebees.org.uk

BA

BEES ABROAD UK Ltd

Bees Abroad

Supporting beekeeping projects overseas

ADMINISTATOR
MRS JULES MOORE
PO BOX 2058
BRISTOL
BS35 9AF
0207 7193 7135
info@beesabroad.org.uk

Bees Abroad is a UK-registered charity (No 1108464) which was established in 1999. Its principle aim is the relief of poverty in the developing world using beekeeping and associated skills as a tool of individual, group and community empowerment for poverty alleviationand to provide sustainable income. Beekeeping is a valuable tool as it is socially and culturally acceptable for both genders across a wide age range.It can cost very little to set up a beekeeping operation, which will deliver benefits for income, education, health, environment and community. Beekeeping and its associated skills deliver access to gainful self-emplyment for poor and disadvantaged groups. This enables them to recover social status, improve social interactions, obtain income and aquire new skills to build the confidence to represent their own interests. Bees Abroad receives a high volume of direct appeals for assistance from groups all over the world. In practice, it acheives its aims through a volunteer network of supporters, committee members and project managers.Bees Abroad takes care to ensure that its projects are sustainable and not dependent on constant external input. This is done by supporting community group initiatives, setting up village-based field extension services, running training courses for beekeeping trainers and financing local trainers' wages. All Bees Abroad projects are designed to become self-financing after a defined time period, usually 2-3 years, but sometimes longer. Its first two projects in Nepal and Cameroon now employ 42 beekeeper trainers and involve many more. It currently has projects either running or seeking funding in Malawi, Kenya, Ghana, Nepal, Uganda and Nigeria.

Our committee is almost entirely run by volunteers, who are all beekeepers. Volunteers and members currently undertake all activities, including fundraising, though a part-time administrator is employed for one day a week. We also arrange Beekeeping Holidays to variety of locations, including Chile, Cameroon and Kenya.

For more details of what we do and how you can help, you can contact Mrs Jules Moore the Administrator, Bees Abroad. Membership costs £15.00 per annum.

BFD

✉ ☎

BEES for DEVELOPMENT TRUST

supporting beekeepers in developing countries

www.beesfordevelopment.org

We always need more help and skills –
please contact us if you might like to be involved.

Bees *for* Development
1 Agincourt Street
Monmouth
NP25 3DZ
info@beesfordevelopment.org
01600 714848
UK registered Charity
No 1078803

YOU CAN HELP US BY:

- **Sponsoring** a Journal for a beekeeper working in a poor country
- **Making** a gift of a Resource Box for a training in a school or project
- **Giving** a donation
- **Joining** one of our *Beekeepers' Safaris*
- **Buying** from our shop in Monmouth or from our on-line store
- **Attending** one of our *Courses*
- **Offering** your skills to work with us as a volunteer
- **Ensuring** that your group or organisation knows about our work, and supports us if possible
- **Helping** us to present our work at events

We are a professional organisation, working in the beekeeping development sector for 20 years. We are respected and trusted by beekeepers world-wide.

Worldwide

- Providing training and information materials for community groups to improve their knowledge of beekeeping and business
- Publishing **Bees for Development Journal** keeping remote beekeepers in 130 countries up to date with news, practical advice, and events
- Maintaining a large resource base for the sector, available on-line

International Projects

With specific communities and partners we work to increase beekeeping incomes

- Cameroon – better processing equipment to raise quality of saleable produce
- Ethiopia – strengthening market chains and training trainers
- Kyrgyz Republic – helping to solve land use conflicts for beekeepers
- Uganda – enabling beekeepers to access good markets

Shop in Monmouth

All proceeds from sales go to support our charitable work. We sell

- A wide range of local honeys and bee products, African honey
- Beekeeping equipment
- bee-related gifts, cards and books

Information Gallery in Monmouth

- Learn about our work overseas
- See a range of unusual bee hives
- Find out about courses and events

Spring 2013
Convention

Harper Adams University College
Newport Shropshire TF10 8NB

Friday 12 – Sunday 14 April 2013

Three days of lectures, workshops and seminars
featuring leading experts in their field

Friday 12 April – **Members Day**

Saturday 13 April
Public Day and Trade Show

Sunday 14 April – **Education Day**

Further information will appear in the
beekeeping press and on the BBKA web-site in due course.

Tickets on sale – January 2013
On site accommodation bookings open – January 2013

Make a note in your diary!

General enquiries: Tim Lovett tjl@dermapharm.co.uk

Trade Show enquiries: John Hayward
jvhayward@suffolkonline.net

BBKA
✉ ☎

BRITISH BEEKEEPERS' ASSOCIATION www.bbka.org.uk

COMMITTEES OF THE EXECUTIVE AND SECRETARIES

FINANCE
This team of Trustees reviews & agrees all budgets, handles all investment matters, finalising insurance policies and sets proposals relating to capitation.

Governance
Primary areas of responsibility are to ensure that we adhere to Charity Commission rules, that we operate within the constitution in addition to ensuring that our Trustees act in the best interests of the BBKA and it's members.

Operations & Membership Services
Headed by the Vice Chairman this team ensure that all Membership Services are administered effectively and on time and that the organisation operates efficiently. It also acts as a co-ordinator for all external fundraising.

Technical & Environmental
All technical issues and their their potential impact on bees and beekeeping are assessed and monitored within this team. All research projects are reviewed and recommendations made by Technical & Environmental group.

Public Affairs
Whether it be government liason, both UK & EU, or press activity this comes from the Public Affairs team.

BBKA

✉ ☎

PRESIDENT
MARTIN TOVEY
martin.tovey@bbka.org.uk

CHAIRMAN
DR DAVID ASTON
david.aston@bbka.org.uk

VICE CHAIRMAN
DOUG BROWN
doug.brown@bbka.org.uk

TREASURER
michael.sheasby@bbka.org.uk

All enquiries should be made to BBKA Press Officer:
gill.maclean@bbka.org.uk

Education & Training

The development of information from practical guidance notes, advisory leaflets, training materials while also undertaking it's own educational initiatives in support of improving the knowledge and skills of beekeepers at all levels. Education & Training liase with the Examination Board to develop training materials to support Association tutors with products such as the Course in a Case.

Examination Board

The BBKA examination board provide a structured range of examinations fulfilling the needs of all beekeepers from Junior Certificate to Master Beekeeper. The board are responsible for all matters relating to the syllabus, content and assessment and operate independently of the BBKA board of Trustees. Where Associations have no Examinations Secretary the Association Secretary deals with examinations. To help future candidates it is suggested that Associations without an Examination Secretary appoint one. Associations are responsible for arranging a suitable room for the written examinations and recommending an invigilator.
Contact Val Frances, Exam Board Secretary
Email: val.frances@bbka.org.uk Tel 01226 286341

Insurance

Members of BBKA, Area Associations and officials are indemnified against claims for Public Liability to a limit of £10million, Product Liability to a limit of £10 million, Professional Indemnity to a limit of £2 million relating to their beekeeping activities. BBKA Association Officer and Trustee liability insurance also applies to a limit of £10 million. Each new claim carries an excess payable by the member.
An 'All Risks' policy is available to both individuals and Associations, to cover the loss or damage of property & equipment. Details are available via www.bbka.org.uk or the main office: 02476 696679

Publications
- BBKA News is issued monthly free to all members of the BBKA, featuring articles about bees, beekeeping and the other associated articles of interest. Editorial: editorial@bbkanew.org.uk Advertising: advertising@bbkanews.org.uk
- BBKA Year Book is published each Spring and is for Association use and reference.It contains detailed information on the BBKA including useful reference tools such as a directory of Lecturers and Demonstrators.
- Members Handbook is published annually and sent to new members
- BBKA Introduction to Beekeeping

BBKA Website - www.bbka.org.uk
The BBKA Website contains technical information, is easy to navigate and supports both beekeepers and the general public. You can download publications, find help and advice in the discussion forums, purchase merchandise, learn about Bees, use the Bees4kids section, download BBKA exam application forms and the exam syllabus. Within the Members Only area, specific insurance downloads and other member only information is available. Associations beekeeping events are promoted.
A Swarm Collector database is included within the site enabling the general public with a direct link to a local swarm collector.

Events
Area and local associations attend and exhibit at various events within their local throughout the year while the BBKA supports selected national shows. Whether it be village fete or national exhibition these events continue to provide a vital service for the dissemination of knowledge.

BBKA Spring Convention
Held in April every year this is a firmly established major beekeeping event. Lectures and Workshops are

staged over 3 days with a one day trade exhibition on the Saturday. Both Friday and Sunday are member only days which are ticketed.

Slide and Video Library
A comprehensive catalogue of the BBKA Slide and VHS Video Library is available from Bridget Knutson, BBKA Slide & Video Librarian. The slides are 35mm, both colour and B&W; some are supported with lecture notes. Many sets are now available as Power Point presentations, please enquire. We are always pleased to receive suggestions for new additions to our range and for donations of items to improve the service.
Please contact Mrs Bridget Knutson E mail: bridget_knutson@yahoo.co.uk

Subscriptions & Membership Fees
Individual Membership of the BBKA is £33 per annum, for an Overseas Member the fee is £25.00. All other membership is via local associations.
Friends Membership is also available via www.bbka.org.uk

Exam Board Footnote
Where Associations have no Examinations Secretary the Association Secretary deals with examinations. To help future candidates it is suggested that Associations without an Examination Secretary appoint one. Associations are responsible for arranging a suitable room for the written examinations and recommending an invigilator.
If you live in an area without a nominated Exam Secretary, you should contact Val Frances, Exam Board Secretary Email: val.frances@bbka.org.uk Tel 01226 286341

BBKA Enterprises
BBKA Enterprises Ltd is a private company, limited by guarantee with all profits from the trading activities being donated to the BBKA. Via the BBKA online shop a range of beekeeping, corporate and related items, specially selected books, gifts, travel items and educational materials are available.
Visit the www.bbka.org.uk/shop or Tel: 02476 696679.

BBKA

AREA ASSOCIATION SECRETARIES

AVON, Jane Godwin
Chatleigh house
6 Warminster Road
Limley Stoke
Bath
BA2 7GD
01225 723292
janegoodwin@macace.net
BERKSHIRE, Martin Moore,
19 Armour Hill
Tilehurst
READING
RG31 6JP
01189677386
07729620286
secretary.berksbees@
uwclub.net
BOURNEMOUTH
Mr A Curry
40 Lacy Drive,
Wimborne, BW21 1DG
01202 840993
andrew@curry@virgin.net
BUCKS, Dr Beulah Cullen
26 Sweetcroft Lane,
Uxbridge, UB10 9LD,
01895 234704
beulah.cullen@virgin.net
CAMBRIDGESHIRE
Mrs Judith Evans MBE
7 The Furlongs
Needingworth
St Ives
Cambridgeshire PE27 4TX
01480 461203
judith@evans.cambnset.co.uk

CHESHIRE, M.F. Haynes
98, Gatley Road, Gatley,
Cheadle,
Cheshire SK8 4AB
0161 491 2382
thesecretary@cheshire-bka.
co.uk
CHESTERFIELD, Robin Bagnall
21 Ramper Avenue,
Clowne, Chesterfield
Derbyshire S43 4UD
01246 570545
ancient.mariner74-79@
virgin.net
CORNWALL, Julia Cooper
Whistow Farm, Lanlivery,
Bodmin, PL30 5DE
01208 872865
julia.i.cooper@btinternet.com
CORNWALL WEST
Mrs Berenice Robbins
Dr Anne McQuade,
5 Trevellan Road, Mylor
Bridge, Falmouth
Cornwall, TR11 5NE
01326 373749
CUMBRIA, Stephen C Barnes
8 Albemarle Street
Cockermouth
Cumbria CA13 0BG
01900 824872
braithwaitebees@sky.com
DERBYSHIRE, M J Cross
Harlestone, Beggarswell
Wood, Ambergate
Derbyshire DE56 2HF
01773 852772
crosssk@btinternet.com

DEVON, Andrew Kyle
62 Bicton Street,
Exmouth EX8 2RU
01395 263509
andrewkyle@tiscali.co.uk
DORSET, Mrs Ruth Homer
5, Malters Cottage,
Litton Cheney,
Dorchester DT2 9AE
beekeepers@hotmail.com
DOVER & DISTRICT
Mrs Maggie Harrowell
4 Harton Cottages, Ashley,
Dover, CT15 5HS
01304 821208
the.harrowells@btinternet.com
DURHAM, John Metson
7 Sidegate, Durham City,
Durham
0191 384 5170
ESSEX, Mrs Pat Allen
8 Frank's Cottages
St Mary's Lane
Upminster RM14 3NU
01708 220897
pat.allen@btconnect.com
GLOUCESTERSHIRE
Marie Toman
Oak Cottage,
Stoulgrove Lane, Woodcroft,
Chepstow, Mon., NP16 7QE.
01291 620345
marietoman@btconnect.com
GWENT
Mrs J Bromley
Ty Hir, Monmouth Road
Raglan, Usk. NP15 2ET
01291 690331
bromleyjan@hotmail.com

BBKA

HAMPSHIRE, Mrs P Barker
Brookdean, Hillbrow
Liss, Hampshire GU33 7PT
01730 895368
H'GATE & RIPON
Mr Frank Ward
19 High Street
Starbeck, Harrogate
HG2 7NS
01423 880266
HEREFORDSHIRE
Mrs. Wendy Cummins
Brook Cottage
Whitbourne
Worcestershire WR6 5RT
01886 821485
jerryandwendy@btinternet.com
HERTFORDSHIRE
Luke Adams
53 Park Street Lane
Park Street
St. Albans
Hertfordshire AL2 2JA
01442 843 779
luke.skywalker@virgin.net
HUNTINGDON,
Nick Steiger
Bull Cottage, Main Street,
Upton, Huntingdon, Cambs,
PE28 5YB
01480 891935
n.steiger@btinternet.com
ISLE OF MAN,
Janet Thompson
Cott ny Greiney, The Smelt,
Beach Road, Port St Mary,
Isle of Man, IM9 5NF
01624 835524
jthompson@manx.net

ISLE OF WHITE,
Mrs. Mary Case
Limerstone Farm
Limerstone,
Newport
Isle of White PO404AB
01983 759510
KENDAL &
SOUTH WESTMORLAND
Roger Blocksidge
Castle Garden Cottage
Aynam Road, Kendal
Cumbria LA9 7DE
01539 734436
tinky_winky@hotmail.com
KENT, John D. Hendrie
26 Coldharbour Lane
Hildenborough, Tonbridge
Kent TN11 9JT
01732 833894
jdh@bbka.freeserve.co.uk
LANCASHIRE & NORTH WEST
Martin Smith
137 Blaguegate Lane,
Lathom
Skelmersdale
Wigan WN8 8TX
05601 484388
ormskirk_beekeepers@
hotmail.com
LINCOLNSHIRE
Mrs. Celia Smith
Brookfield, Moor Town Road,
Nettleton LN7 6HX
01472 851165
LONDON
Nikki Vane
601 Alaska, 61 Grange Road
London SE1 3BB
07909 964986
sec@lbka.org.uk

LUDLOW & DIST
Andy M Vanderbrook
The Old Forge
Baveney Wood
Cleobury Mortimer
Kidderminster
DY14 8JD
01299 841379
andy.vanderhook@
care4free.net
MANCH. & DIST, Mrs. M. Bohme
54 Dunster Drive, Flixton
Manchester M41 6WR
0161 747 7292
MEDWAY Mrs. M. Pines
26 Lapwing Road
Isle of Grain, Rochester
ME3 0EB
01634 272252
MIDDLESEX, Mrs. J.V. Telfer
Midwood House
Elm Park Road
Pinner HA5 3LH
020 8868 3494
jvtelfer@hotmail.com
NEWCASTLE & DISTRICT,
Mr D Varty
Cragside, Dipton
Stan DH9 9EL
01207 570229
NORFOLK, Mrs H Coppwaite
1 The Maltings, Millgate
Aylsham, Norwich
NR11 6GX
01263 734682
NORFOLK WEST & KINGS LYNN
Mrs Irene Laws
16 Pine Road,
South Wootton
King's Lynn PE30 3JP
01553 671312

NORTHAMPTONSHIRE
Mrs Ruth Stewart
17 Leys Avenue, Rothwell,
Kettering,
Northants, NN14 6JF
01536 507293
rstewart@euramax.co.uk

NORTHUMBERLAND
Mr Ben Hopkinson
11 Watershaugh Rd
Warkworth, Northumberland
NE65 0TT
01665 714213
benhopkinson2436@waitrose.com

NOTTINGHAMSHIRE
M. Jordan
29 Crow Park Avenue
Sutton on Trent
Nr Newark NG23 6QG
01636 821613
mauricejordan11@btinternet.com

OXFORDSHIRE, Mr Mark Lynch
71 Millwood End Long
Hanborough
Oxfordshire, OX 29 8BP
01993 883266
marka_lynch@hotmail.com

PETERBOROUGH & DISTRICT
P George Newton
65 Queen Street, Yaxley
Peterborough PE7 7JE
01733 243349

ROSELAND BEEKEEPING GROUP
Rose Hardisty
Menagwins Cottage,
Pentewan Road, St Austell,
Cornwall PL26 7AN
01726 74101
roselandbee@tiscali.co.uk

SEDBURGH
Jane Callus-Whitton
Harren House, Woodman
Lane, Cowan Bridge,
Carnforth, LA6 2HT
01524 272004

SHROPSHIRE
Mrs Penny Carkeet-James
Upper Dumble Holes
Westbury
Shrewsbury SY5 9HE
01743 791081

SHROPSHIRE NORTH
Mrs Jo Schup
Fields Farm, Malt Kiln Lane
Dobsons Bridge,
Whixall SY13 2QL
01948 720731

SOMERSET,
Mrs S Perkins
Tengore House,
Tengore Lane, Langport
Somerset TA10 9JL
01458 250095
bernieperkns.tengor@tiscali.co.uk

STAFFORDSHIRE NORTH
Janey Hayward
90 Ostler's Lane
Cheddleton ST13 7HS
01538 361048

STAFFORDSHIRE SOUTH
Mr Steve Halford
46 Lincoln Hill Telford
TF8 7QA
01952 432031
stevehal@tiscali.co.uk

STRATFORD-ON-AVON
Michael Osborne
Oak Lodge, King's Lane
Snitterfield
Stratford-upon-Avon
Warwickshire CV37 0RB
01789 731745
mjroosborne@btinternet.com

SUFFOLK,
Ian McQueen
643 Foxhall Road
Ipswich IP3 8NE
01473 420187
jackie.mcqueen@ntlworld.com

SURREY, Mrs Sandra Rick
19 Kenwood Drive,
Walton-on-Thames
Surrey. KT12 5AU
01932 244 326
rickwoodsbka@googlemail.com

SUSSEX, Mrs Moyra Davidson
Gainsborough Cottage, Stunts
Green, Hertsmonceux,
East Sussex, BN27 4PN
01323 831 650
secretary@sussexbee.org.uk

SUSSEX WEST,
Mr John Glover
Fletchings Hollow
Vicarage Hill, Loxwood,
West Sussex RH14 0RJ
01403 751 899
glover.fletchershollow@googlemail.com

THANET,
Mrs R Pearce
Summerfield Cottage
Summerfield
Woodnesborough
Nr Sandwich CT13 0EW
01304 614789

BBKA

TWICKENHAM & THAMES VALLEY
(MOLE APIARY CLUB)
Mrs Sarah Crofton
11 Wellesley Avenue
London TW3 2PB
0208 222 8216
WARWICKSHIRE,
Theresa Simkin
87 Kineton Green Road
Solihull, B92 7DT
secretary@
warwickshirebeekeepers.org.uk

WILTSHIRE,
Ruth Woodhouse
Sandridge Tower
Bromham
Devizes
SN15 2JN
01225 705382
sandridgetower@aol.com
WORCESTERSHIRE
Mr Chris Broad,
Upper Gambolds Farm,
Upper Gambolds Lane,
Bromsgrove
Worcestershire
B60 3HD
01527 872448
chrisbroad1964@btinternet.
com

WYE VALLEY, Mrs S Wenczek
Hopleys, Bearwood
Leominster HR6 8EQ
01544 388302
YORKSHIRE, Brian Latham
111 Woodland Road
Whitkirk, Leeds
LS15 7DN
0113 264 3436
chrisbroad1964@btinternet.
com

ASSOCIATION EXAMINATION SECRETARIES

AVON, Position Vacant
Please contact
Hon. General Secretary
Julie Young
01179 372 156
BERKSHIRE,
Mrs Rosemary Bayliss
Norbury, Coppid Beech Hill,
Binfield, Berkshire.
RG42 4BS
01344 421747
BOURNEMOUTH, Mrs. M. Davies
57 Leybourne Avenue
Ensbury Park
Bournemouth
Dorset BH10 6ES
01202 526077

BUCKS, John Chudley
Orchard Lea, Oxford Street
Lea Common
Great Missenden HP16 9JT
01494 837544
jlchudley@tiscali.co.uk
CHESHIRE
Graham Royle NDB,
7, Symondley Road,
Sutton,
Macclesfield. SK11 0HT
01260 252 042
CORNWALL
Mrs. Susan Malcolm
Fig Tree, 333 New Road
Saltash, Cornwall
PL12 6HL
01752 845496

DEVON, Roger Lacey
Gatchell House
Toadpit Lane, Ottery St Mary
Devon EX11 1TR
01404 811733
devonbees@pobox.com
DORSET, K.G.Bishop
72 Alexandra Road
Bridport DT6 5AL
01308 425479
DURHAM, G. Eames
23 Lancashire Drive
Belmont, Durham,
DH1 2DE
01913 845220
george.eames@durham.ac.uk
ESSEX, Pat Allan

8, Frank's Cottages
St. Mary's Lane
Upminster, RM14 3NU
pat.allen@btconnect.com
GLOUCESTERSHIRE
Bernard Danvers
120a Ruspidge Road
Cinderford,
Gloucestershire
GL143AG
01594 825063
GWENT, Mrs J Bromley
Ty Hir, Monmouth Road
Raglan, Usk. NP15 2ET
01291 690331
bromleyjan@hotmail.com
HAMPSHIRE, Mrs Peggy Mason
37 Springford Crescent
Lordswood,
SO16 5LF
023 8077 7705
H'GATE & RIPON, Peter Ross
The Wheelhouse, The Green,
Old Scriven, Knaresborough
HG5 9EA, 01423 866565,
pjeross@btinternet.com
HEREFORDSHIRE, Len J. Dixon
The Square, Titley,
Kington
Herefordshire HR5 3RG
01544 230884
beeline2ljd@yahoo.co.uk
HERTFORDSHIRE, R. E. A. Dart-ington
15 Benslow Lane
Hitchin SG4 9RE
01462 450707
gray.dartington@dial.pipex.com
ISLE OF WIGHT, Mrs M. Case
Limerstone Farm,
Limerstone, Newport,
Isle of Wight, PO30 4AB
01983 740223
gcase90337@aol.com

KENT, P. F. W. Hutton
22 Good Station Road
Tunbridge Wells,
TN1 2DB
01892 530688
LANCASHIRE & NW
Edward Hill
3 Sandy Lane, Aughton
Ormskirk
L39 6SL
01695 423137
LEICESTERSHIRE & RUTLAND
Brian Cramp
2 Woodland Drive, Groby
Leicester
LE6 0BQ
01162 876879
LINCOLNSHIRE, R. J. B. Hickling
Linden Lea, Sandbraes
Lane, Caistor, LN7 6SB
01472 851473
MIDDLESEX
Mrs Jo V Telfer
Midwood House
Elm Park Road, Pinner
Middlesex HA5 3LH
020 8868 3494
e-mail, jvtelfer@hotmail.com
NOTTINGHAMSHIRE
Dr Glyn D Flowerdew
Knight Cross Cottage
Newstead Abbey Park
Ravenshead
Nottinghamshire NG15 8GE
01623 792812
OXFORDSHIRE, Terry. Thomas
4 Kirk Close
Oxford, OX2 8JN
01865 558679
PETERBOROUGH, P. G .Newton
65 Queen Street, Yaxley
Peterborough PE7 3JE
01733 243349

SHROPSHIRE NORTH
Paul Curtis
1 Hammer Close
Overton-on-Dee, Wrexham
Clwyd LL13 0LD0
01691 624296
SOMERSET, Mrs Angela Bache
Greenway House
Badgers Cross
Somerton TA11 7JB
Tel 01458 273149
STAFFS.Nth Dr. Nick C Mawby
Glenwood, Wood Lane
Longsdon,
Stoke on Trent ST9 9QB
01538 387506
info@northstaffsbees.org.uk
STAFFS. SOUTH
Tony Burton
96 Weeping Cross, Stafford,
Staffordshire. ST9 9QB
01538 399322

BBKA

SUFFOLK, Mr Ian McQueen
643 Foxhall Road, Ipswich,
Suffolk, IP3 8NE
01473 420187

SURREY, Mrs. A. Gill
143 Smallfield Road
Horley, RH6 9LR
01293 784161

SUSSEX, Nigel Champion
45 Ridgeway,
Hurst Green
Etchingham
East Sussex TN19 7PJ
01580 860379

SUSSEX WEST
Mrs A. S. Gibson-Poole
Mont Dore, West Hill
High Salvington
Worthing, BN13 3BZ
01903 260914

TWICKENHAM, Chris Deaves
12 Chatsworth Crescent
Hounslow,
Middlesex
TW3 2PB
0208 5682869
e-mail,
c-deavs@compuserve.com

WARWICKSHIRE, P.D. Lishman
Aston Farm House
Newtown Lane
Shustoke, ColeshillB46 2SD
01676 540411

WILTSHIRE, John Troke
The Lythe
Hop Gardens
Whiteparish, Salisbury,
Wiltshire SP5 2SS
01373 822892

WORCESTERSHIRE, D.P. Friel
17 Tennal Rd, Harborne
Birmingham, B32 2JD
0121 427 1211

YORKSHIRE, Brian Latham
Tel: 0113 264 3436
Mob: 07765842766
brian.latham@ntlworld.com

Where Associations have no Examinations Secretary the Association Secretary deals with examinations. To help future candidates it is suggested that Associations without an Examination Secretary appoint one. Associations are responsible for arranging a suitable room for the written examinations and recommending an invigilator.

If you live in an area without a nominated Exam Secretary, you should contact Mrs Val Frances, 39 Beevor Lane, Gawber, Barnsley, S75 2RP Tel 01226 286341. e-mail, valfrances@blueyonder.co.uk

BBKA
✉ ☎

HOLDERS OF THE BBKA SENIOR JUDGES CERTIFICATE

ASHLEY, Mr. T. E.
Meadow Cottage
Elton Lane, Winterley
Sandbach
Cheshire CW11 4TN
BADGER, M.J , MBE
14 Thorn Lane
Leeds, LS8 1NN
BLACKBURN, Mrs. H.M
15 Highdown Hill Road
Emmer Green
Reading RG4 8QR
BROWN, Mrs. V
BUCKLE, M.J
The Little House
Newton Blossomville
Bedford MK43 8AS
01234 881262
martin@newtonbee.fsnet.
co.uk
CAPENER, Rev. H.F.
1 Baldric Road
Folkestone CT20 2NR
COLLINS, G.M. , NDB
72 Tatenhill Gardens
Doncaster DN4 6TL
COOPER, Miss R.M
10 Gaskells End
Tokers Green
Reading RG4 9EW
DAVIES, Mrs. M
57 Leybourne Avenue
Ensbury Park
Bournemouth BH10 6HE

DIAPER, B
B Diaper
57 Marfield Close
Walmley
West Midlands
0121 313 3112
or 07711 456932
DICKSON, Ms. F
Didlington Manor
Didlington, Thetford
Norfolk IP26 5AT
DUFFIN, J.M
Upper Hurst
Salisbury Road, Blashford
Ringwood
Hampshire BH24 3PB
01425 474552
FIELDING, L.G
Linley, Station Road
Lichfield WS13 6HZ
MacGIOLLA CODA, M.C.
Glengarra Wood, Burncourt
Cahir, Co. Tipperary
Republic of Ireland
McCORMICK, E.
14 Akers Lane, Eccleston St.
Helens, Lancs WA10 4QL
MOXON, G
9 Savery Street
Southcoates Lane
Hull HU9 3BG

ORTON J
Occupation Road, Sibson
Nuneaton CV13 6LD
ROUNCE J.N , NDB
4 Scarborough Road
Great Walsingham
NR22 6AB
SALTER T.A , MBE
44 Edward Road, Clevedon
North Somerset BS21 7DT
SYMES, C.J
189 Marlow Bottom Road
Marlow SL7 3PL
TAYLOR, A.J
The Old Pyke Cottage
Hethelpit Cross, Staunton
GL19 3QJ
WILLIAMS, M
Tincurry, Cahir,
Co Tipperary, Eire
YOUNG, M
Mileaway, Carnreagh
Hillsborough,
Northern Island BT26 6LJ

BIBBA

BEE IMPROVEMENT & BEE BREEDERS' ASSOCIATION

www.bibba.com

SECRETARY
Dinah Sweet
Craig Fawr Lodge
Caerphilly
CF83 1NF.
02920 869242
dinah@dinahsweet.com

MEMBERSHIP SECRETARY
Enid Brown
Milton House
Main Street
Scotlandwell
Kinross-shire
KY13 9JA
01592 840582
honeybees@onetel.com

SALES SECRETARY
John Hendrie
26 Coldharbour Lane
Hildenborough
Tonbridge
Kent
TN11 9JT
sales@bibba.com

BIBBA is an organisation devoted to encouraging beekeepers to breed native bees. The bee more suited to our environmental circumstances than other sub species. BIBBA's aims are publicised through books, workshops, lectures and conferences.

BIBBA also co-operates with worldwide Beekeeping and breeding groups interested in conserving and improving their own native bees.

Breeding techniques advocated include:
• Assessment of colonies by observation, recording certain criteria on standard record cards.
• Determination and purity of sub species by measurement of morphometric characters and mitrochondial DNA.
• Use of mini nucs for the mating of queens economically

BIBBA Publications include:
• The Honeybees of the British Isles by Beowulf Cooper
• Breeding Techniques and Selection for Breeding of the Honeybee by Prof. F. Ruttner
• The Dark European Honey Bee by Prof. F. Ruttner, Rev. Eric Milner and John Dews
• Breeding Better Bees using Simple Modern Methods by John E. Dews and Rev.Eric Milner
• Better Beginnings for Beekeepers by Adrian Waring - second edition.

BIBBA encourages the formation of Bee Breeding Groups, and the sharing of knowledge between groups by the provision of genetic material.
Look out for Queen Rearing events in the bee press and on www.bibba.com.

CBDBBRT

✉ ☎

THE C.B. DENNIS BRITISH BEEKEEPERS' RESEARCH TRUST

REGISTERED CHARITY NO. 328685

Aims

This Charitable Trust was established in 1990 through the generosity of Mr C.B. Dennis. It aims to support research at institutions or by individuals and to encourage young scientists through the provision of grants for agreed projects and broader bee science that benefit bees and beekeeping in Britain.

Awards

The Trust is administered by a group of Trustees, most of whom have relevant scientific, environmental or ecological expertise that helps to ensure that work funded by the Trust is properly evaluated and provides the greatest possible advantage for bees.

Since its inception the Trust has funded work on a wide range of topics related to both honey bees and other bees. Grants have been made, for example, for work on: bee pathogens, pollination and broader ecological concepts. Work is not limited to the UK and research at European universities and laboratories may also be funded.

Young scientists have been supported through studentships and applications for such research bursaries are invited.

Donations

The Trust is pleased to acknowledge the loyal support it already receives from several local beekeeping associations and many individuals. All donations, however small, will be added to the invested capital and bee research in Britain will benefit from the income in perpetuity.

Full details of the activities of the Trust, outputs of the research funded and grant application forms can be obtained from www.cbdennistrust.org.uk

CABK

✉ ☎

THE CENTRAL ASSOCIATION OF BEEKEEPERS

www.cabk.org.uk

SECRETARY, Pat Allen
8 Frank's Cottages
St Mary's Lane
Upminster, RM14 3NU
pat.allen@btconnect.com

PRESIDENT, Prof. R.S. Pickard
Consumer's Association
pickard.r@btopenworld.com

TREASURER, John Hendrie
26 Coldharbour Lane
Hildeborough
Tonbridge, TN11 9JT
bibba26@talktalk.net

PROGRAMME SECRETARY
Pam Hunter
Burnthouse
Burnthouse Lane
Cowfold, Horsham
RH13 8DH
pamhunter@burnthouse.org.uk

EDITOR, Pat Allen
8, Frank's Cottages
St. Mary's Lane
Upminster, RM14 3NU

SALES AND DISTRIBUTION,
Margaret Thomas
Tighnabruaich,
Taybridge Terrace,
Aberfeldy, Perthshire
PH15 2BS
zyzythomas@waitrose.com
01887 829710

The Central Association of Beekeepers in its present form dates from the time of the reorganisation of the British Beekeepers' Association in 1945. The BBKA was originally made up of private members only. However as County Associations were formed they applied for affiliation and were later permitted to send delegates to meetings of the Central Association, as the private members were then known. This arrangement became unsatisfactory as the voting power of the Central Association greatly outnumbered that of the County Associations and so in 1945 a new Constitution was drawn up whereby the Council comprised Delegates from the Counties and Specialist Member Associations. The private members then formed themselves into a Specialist Member Association with the designation 'The Central Association of the British Beekeepers' Association'; this was later shortened to its present style.

The Association was able to devote itself to its own particular aims, to promote interest in current thought and findings about beekeeping and aspects of entomology related to honey-bees and other social insects. Lectures given by scientists and other specialists are arranged, printed and circulated to members, as has been done since 1879.

An annual Spring Conference is held in London and an Autumn Conference in the Midlands. In addition, a lecture is presented at the Annual General Meeting and at the Social Evening held during the National Honey Show. The subscription is £10.00 per annum, £12.00 for dual membership (one copy only of publications).

BEE MAGS

✉ ☎

COUNTY BEEKEEPING MAGAZINES AND NEWSLETTERS

AVON, Ms Julie Young
1 Church Cottages
Abson Road, Abson Wick
Bristol, BS30 5TT
0117 937 2156
julieyoung@btinternet.
com

BEDFORDSHIRE, Sue Lang
154a Lower Shelton
Road, Upper Shelton
Marston Moretaine
Beds, MK43 0LS
01234 764180
07879 848550
bedfordshirehoney@
hotmail.co.uk

BERKSHIRE, Ron Crocker
25 Ship Lake Bottom
Peppard Common
Oxon RG9 5HH

BRADFORD
Bill Cadmore,
104 Hall Lane, Horsforth
Leeds LS18 5JG
0113 216 0482
bill.cadmore@
ntlworld.com

CAMBRIDGESHIRE,
Mr. Chris Evans
7 The Furlongs,
Needingworth,
St. Ives,
Cambs. PE27 4TX

CHESHIRE,Pete Sutcliffe
2 Hatfield Court
Holmes Chapel,
Cheshire, CW4 7HP
h.p.sutcliffe@googlemail.
com.

CHESTERFIELD & DISTRICT
Mrs Margaret Edge
4 Cinder Hill,
Shireoaks,
Worksop, S81 8NR

CORNWALL, Gillian Searle
6 Harleigh Road, Bodmin
Cornwall PL31 1AQ

CUMBRIA, Dave Bates
Greenfield House
Low Green
Temple Sowerby
Penrith CA10 1SD

DERBYSHIRE, Mrs. M. Cowley
14 Montpelier, Quorndon
Derby DE22 5JW

DEVON, Glyn R Davies
Landscore
Eastern Rd, Ashburton
Devon TQ13 7AR
01364 652640
landscore@eclipse.co.uk

DORSET, Mrs L Gasson
The White House,
Candy's Lane,
Shilllingstone. DT11 0SF
01258 861690

DURHAM, George Eames
11, Sharon Avenue,
Kelloe, Durham DH6 4NE
07970 926250
beeseames@btinternet.
com

ESSEX, Pat Allen
8 Franks Cottages
St Mary's Lane
Upminster RM14 3NU

GLOUCESTERSHIRE, Mrs A Ellis
19 Whaddon Road
Cheltenham
Gloucestershire GL52 5LZ

GUERNSEY BKA
Ruth Collins
Colombier House
Torteval
Guernsey GY8 0NF

GWENT, Keith Allen
Pen-y-Lan Cottage,
Far Hill, Trellech,
Monmouth, NP25 4PP
kh.allen@virgin.net

HAMPSHIRE, Dr Helen Harley
Communications Manager
Programme Management
Unit (PMU)
University of
Southampton
Bassett House
Chetwynd Road
Southampton SO16 3TU
023 8059 2804

BEE MAGS

✉ ☎

HEREFORDSHIRE, Lin Hoppé
07974 960 956
lin@thesteppeshereford.
co.uk
HERTFORDSHIRE
Paul Cooper
01279 771231
HUNTINGDON
Wilma Vaughan
Lauriston Copse
Warboys, Huntingdon
Cambs PE28 2US
KENDAL & SOUTH
WESTMORLAND
Roger Blocksidge
20 Fowl Ing Lane
Kendal
Cumbria
LA9 6HB
tinky_winky@
hotmail.com
KENT,John Hendrie
26, Coldharbour
Lane,Hildenborough,
Tonbridge Kent
TN11 9JT
LEEDS BEEKEEPER
Editor, Katey Slater
editor@leeds
beekeepers.org.uk
LEICS. & RUTLAND
Editor, Simon Skerritt
1 Orchard Lane,
Countesthorpe,
Leicestershire LE8 5RB
hyper.julia@gmail.com
LINCOLNSHIRE, P. Raines
Grange Cottage
21 Humberston Av.
Humberstone
Grimsby DN36 4SL

LONDON, Steve Bembow
156 Devon Mansions
Tooley Street
London SE1 2NR
MEDWAY, Rob Smith
Alex Barker
5 Crossway
Rochester
Kent ME1 3DX
01634 406360
alex@alexbarkeronline.
com
MOLE APIARY CLUB,
Dennis Cutler
70 Hurst Road
East Molesey
Surrey KT8 9AG
NEWCASTLE, George Batey
Rift Farm Cottage
Wylam NE41 8BL
NORFOLK, Michael Lancefield
Candlemas House
Fakenham Road
Stanhoe
King's Lynn PE31 8PX
lancefield@aol.com
NOTHAMPTONSHIRE
Roger G Virgo
5 Surfleet Close, Corby
Northamptonshire
NN18 9BG
amellifera@aol.com
NOTTINGHAMSHIRE,
Stuart Ching
122 Marshall Hill Drive
Porchester
Notts NG3 6HW

SOMERSET, Richard Bache
The Annex,
Moorview Farm
Midelney Road
Drayton, Near langport
Somerset, TA10 0LW
newsletter@somerset
beekeepers.org.uk
SUFFOLK, Tony Molesworth
Kizimbani, Bildeston
Road, Combs. IP14 2JZ
tony.molesworth@essex.
businesslink.co.uk
WARWICKSHIRE, Rob Jones
124 Ashfurlong Road
Sutton Coldfield
B75 6EW
0121 378 0562
wbeditor@warwickshire-
beekeepers.org.uk
WILTSHIRE,
Ronald A Hoskins,
10 Larksfield
Covingham Park
Swindon SN3 5AD
WORCESTER,
Martyn Cracknell
Honeylands
Abberton Road
Bishampton
Worcestershire
WR10 2LU
01386 462385
email: martyn@crack-
nellz.freeserve.co.uk
YORKSHIRE
Bill Cadmore,
104 Hall Lane, Horsforth
Leeds LS18 5JG
0113 216 0482
bill.cadmore@
ntlworld.com

CONBA

✉ ☎

CONBA-UK & Ireland
COUNCIL OF NATIONAL BEEKEEPING
ASSOCIATIONS IN THE UNITED
KINGDOM and IRELAND

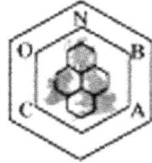

CONBA was established in 1978 to promote the aims and objectives of the national beekeeping associations of England, Scotland, Ulster, Wales and Ireland, and the Bee Farmers Association. Its purpose is to represent the interests of beekeepers' with local, national and international authorities. A representative delegate from each of the member country associations occupies the chair for a period of two years, on a rotational basis.

The council meets twice per year, normally at the Spring Convention and at the National Honey Show in London, with the remaining meeting by rotation in the member association's country. Council business consists of any matters of common interest to all its members. CONBA provides representation of its membership at the European Union (EU) through two specific committees, COPA and COGECA (COPA – Comite des Organisations Professionelles Agricoles de la CEE); (COGECA Comite de la Cooperation Agricole de la CEE); and the Honey Working Party (HWP).

The Honey Working Party meetings are held at Brussels. This committee liases with the European Commission in relation to apicultural matters concerning the member states of the European Union (EU). These matters are subsequently presented to the European Parliament for its consideration, implementation or revision or rejection. The subsequent approval of such matters results in establishing legislation, government support and possible EC funding relating to the practice of apicultural production in the UK through its membership of the EU.

Incorporating the beekeeping organisations of :
England, Channel Islands Isle of Man, Wales, Scotland, Ulster, Ireland and The Bee Farmers Association

SECRETARY David Bancalari (BFA)
Park Farm Barn
Shorthorn Road
Stratton Strawless
Norfolk. NR10 5NX
01603 755105
wiredbrain@btinternet.com

CHAIRMAN, Mervyn Eddie (UBKA)
3b Old Road
Upper Ballinderry
Lisburn, Co. Antrim. BT28 2NJ
0289 265 2580
eddie_mervyn@yahoo.co.uk

VICE CHAIR Michael Gleeson (FIBKA)
Ballinakill
Enfield
Co.Meath, Ireland.
+353 (0)4695 41433
mgglee@eircom.net

HON.TREASURER Martin Tovey (BBKA)
11, Coach Road
Warton, Carnforth
Lancashire. LA5 9PP
01524 730451
martintovey@hotmail.co.uk

139

CONBA

DARG

DEVON APICULTURAL RESEARCH GROUP

D A R G is an independent group of experienced enthusiastic beekeepers whose primary aim is to collect and analyse data on matters of topical interest which may assist their apicultural education and promote the advancement of beekeeping. At their regular meetings, DARG members discuss various topics in open forum, during which they exchange ideas and information from their personal beekeeping knowledge and experience. They also undertake suitable research projects which further the Group's aims.

TOPICS CURRENTLY BEING UNDERTAKEN
- Use of management (mechanical) methods including shook colonies for varroa control.
- Brood cell size in natural comb.
- A survey of useful bee plants, shrubs and trees in the South West.
- Drone movement between colonies.

In conjunction with Devon BKA
- Survey of Nosema in the County of Devon.
- Survey of drone lying queens in the County of Devon.

PUBLICATIONS AVAILABLE
- **The Beeway Code.** A common sense guide for beginners to help avoid problems with neighbours and produce a safe and peaceful apiary.
- **Seasonal Management.** A useful aid to planning your work effectively
- **Queen Rearing.** Providing detailed help in rearing new queens in order to promote vigorous colonies.
- **The selection of Apiary sites** full of tips for choosing the right sites for your bees.

CHAIRMAN, Richard Ball
Stoneyford Farmhouse
Colaton Raleigh
Sidmouth
Devon EX10 0HZ
01 395 567 356

HON SECRETARY, Roger Lacey
Gatchell House
Toadpit Lane
Ottery St Mary
EX11 1TR
01404 811 733
devonbees@gmail.com

PUBLICATIONS OFFICER, David Loo
25 Woodlands
Newton-St-Cyres, Exeter
Devon
EX5 5BP
0139 285 1472

TREASURER, Bob Ogden
Pennymoor Cottage
Pennymoor, Tiverton
Deven
EX16 8LJ
01363 866687

All titles cost £2.50 per copy (post free) from the Publications Officer (tel. 01392 851472). Discounts are available for BBKA affiliated Associations **Please contact the Publications Officer for details**

141

EAS

✉ ☎

THE EASTERN APICULTURAL SOCIETY OF NORTH AMERICA
www.easternapiculture.org

Kathy Summers, 623 West Liberty Street, Medina Ohio 44256
Kathy@BeeCulture.com

The Eastern Apicultural Society of North America (www.easternapiculture.org) holds their annual Short Course and Conference each Summer during July or August.

In 2013 EAS will be held at West Chester University near Philadelphia, August 5-9, hosted by the Pennsylvania State Beekeepers Association The week starts with a Short Course with an advanced level and a beginning level – something for everyone. This is a fairly intense classroom setting with beeyard activities, focused on increasing your knowledge and ability as a beekeeper. Then on Wednesday the Main Conference begins with lectures and on Thursday and Friday workshops in the afternoon on every topic you can imagine.

We offer workshops on beekeeping and many related activities – some years cooking with honey, soap making, candle making – so much you can't possibly see and hear it all.

In 2014 EAS will be in Kentucky at Eastern KY University in Richmond. So please visit our website – www.easternapiculture.org – watch for details of upcoming conferences. We will have our program, speakers and other activities listed just as soon as it is set.

If you have questions or want more information you can also contact Kathy Summers, EAS Vice Chairman of EAS, kathy@beeculture.com.

FIBKA

THE FEDERATION OF IRISH BEEKEEPERS' ASSOCIATIONS

http://www.irishbeekeeping.ie

ANNUAL SUMMER COURSE

The 2013 Beekeeping Summer Course will take place at the Franciscan College, Gormanston, Co Meath from Sunday 21st to Friday 26th of July. The Guest Speaker will be Flemming Vejsnæs from Denmark.
He is a biologist (M.Sc) with focus on pollination. Since 1991 he is the beekeeping adviser employed by the Danish Beekeepers Association which has a membership of 4,500 members. For the past 20 years the main focus has been on practical varroa-treatment in the organic way, keeping Danish honey free of unwanted medicals. The aim has been to test and convert new varroa control methods to useful organic tools for the practical beekeeper while keeping honey completely free from unwanted residues. He is involved in practical fieldwork and continuous development of beekeeping is important. In the main Danish beekeepers fight varroa in the organic way. Over the years Flemming has travelled to beekeeping congresses abroad, visiting beekeepers in other countries which is important to get new beekeeping inspiration. He is actively involved in the international Coloss network, focusing on the widespread looses of colonies over the last 5-10 years worldwide, which has developed a network in nearly all European countries and is of great benefit for the Danish Beekeepers. During the past 20 years Flemming has been the editor of the Danish beekeeper magazine and is now the webmaster of www.biavl.dk, were there is some English written information.
Keep bees alive, healthy and honey clean is the keyword for modern beekeeping.

For further information and to secure your place, contact the Summer Course Convenors Gerry & Mary Ryan, Deerpark, Dundrum, Co Tipperary (062-71274) or 087-1300751 or 087 9745115, ryansfancy@gmail.com

HON. SECRETARY
Mr. Michael G. Gleeson
Ballinakill Enfield Co. Meath
+ 353 (0)46-9541433 or +
353 (0) 87 6879584
mgglee@eircom.net

PRESIDENT
Mr Seamus Reddy
8 Tower View Park, Kildare,
045 521945
seamusreddy@eircom.net

VICE PRESIDENT
Mr Eamon Magee
222 Lower Kilmacud Road,
Goatstown, Dublin 14.
01-2987611
"mailto:eamonmagee222@
gmail.com

PRO Mr P.McCabe,
"Sherdara"
Beuaulieu Cross
Drogheda, Co. Louth
041 983 6159
philipmccabe@eircom.net

143

FIBKA

✉ ☎

EDITOR,
Ms Mary Montaut
4 Mount Pleasant Villas,
Bray, Co Wicklow.
01-2860497
yram@connect.ie

MANAGER,
Mr David Lee
Scart, Kildorrey,
Co Cork
022-25595
davidleej@eircom.net

TREASURER,
Mrs Bridie Terry
"Ait na Greine", Coolbay
Cloyne, Midleton,Co Cork
0214652141
aitnagreine@gmail.com

EDUCATION OFFICER
Mr Michael G Gleeson
Ballinakill, Enfield,
Co Meath.
046-9541433 &
087-6879584
mgglee@eircom.net

SUMMER COURSE CONVENER
Gerry & Mary Ryan
Deerpark, Dundrum,
Co Tipperary
062 71274
ryansfancy@gmail.com

PUBLICATIONS

- **Having Healthy Honeybees - Published by F.I.B.K.A.**
  Editor John McMullan, Ph.D.
  The aim of this book is to help beekeepers establish healthy honeybee colonies, assess their condition and take appropriate action. Diseases are dealt with in a concise format to improve readability and are referenced to the latest peer-reviewed research. The book emphasises the importance of proper set-up, involving an integrated approach to health management – in effect a preventative system that comes at little extra cost to the beekeeper
  Cost €15 + P & P of €2 each
  Bulk buying available to Associations In packs of 10 or 20 books, available at €12 each + P & P of €10 for packs of 10 or 20. The recommended price is €15 per copy.
  It is highly recommended for those doing the various FIBKA Examinations.
  Available from Mr Michael G Gleeson, Ballinakill, Enfield, Co Meath.
  Tel No (046-9541433) & (087-6879584), mgglee@eircom.net

- **Bees, Hives and Honey** - Published by F.I.B.K.A. -
  Edited by Eddie O'Sullivan.
  This book has been compiled from writings by some of Ireland's most prominent Beekeepers of the present day. It is an instruction book on beekeeping published as a Millennium project and should prove a modern treatise on the craft of beekeeping and its associated products. There are over 200 pages, also many photographs and illustrations. Price €12.70 (Paperback) or €19 (Hardback)
  Available from Eddie O'Sullivan, Phone: 021-4542614, eosbee@indigo.ie

- **The Irish Bee Guide** - Reverend J.D. Digges
  First published in 1904, it was proclaimed as an excellent book on beekeeping. It also won a place as a notable production in the literary context. It eventually ran to sixteen editions and sold seventy-six thousand copies overall. The name was changed in the second issue to The Practical Bee Guide. Now, one hundred years later, a decision has been taken to honour this great work. What better way to do it than to re-issue the book as it was in 1904 when it first entered the literary world. The re-print is an exact

replica of the original first edition. The price per copy is Hardback €30 and Softback €20
Available from Eddie O'Sullivan, Phone: 021-4542614, eosbee@indigo.ie

FIBKA

- **An Beachaire** - The Irish Beekeeper the monthly organ of FIBKA, subscription €25.00 (Irish Republic), £25 Stg (Northern Ireland/Great Britain) post free from The Manager, Mr David Lee, Scart, Kildorrey, Co Cork Tel No (022-25595), davidleej@eircom.net.
Readership of the Journal in Northern Ireland carries third party insurance public liability cover up to €6.500,000 on any one claim and product liability cover up to €6.500,000 on any one claim, on payment of £5.00 Stg extra.

LIBRARY
The library is owned and controlled by FIBKA. It contains very many valuable books ancient and modern, available to members for return postage only. The Librarian is Jim Ryan, Innisfail, Kickham Street, Thurles, Co Tipperary. jimbee1@eircom.net

EDUCATION
The Federation of Irish Beekeepers' Associations (FIBKA) examination system is run by the Education Officer under the direction of the Examination Board; the Board which is made up of members from the FIBKA and the Ulster Beekeepers' Association (UBKA) is appointed by the Executive Council of the FIBKA.
There are seven levels of examination: Preliminary, Intermediate, Senior, Lecturer and Honey Judge Examinations are held during the Summer Course at Gormanston and Preliminary and Intermediate examinations are also held at Provincial Centres.
The Lecturer's examination takes place in the presence of three Examiners, one of whom is the invited Senior Gormanston Summer Course lecturer and also acts as the Extern Examiner.
The Intermediate Proficiency Apiary Practical Examination, the Practical Beemasters Examination and the Apiary Practical component of the Senior Examination are arranged by the Education Officer and take place in the candidate's own apiary during the beekeeping season and are conducted by two Examiners.
The seven levels of examinations for proficiency certificates and their eligibility requirements are as follows:
Preliminary:
For beginners - no prerequisites.
Intermediate:
The Preliminary Certificate of the FIBKA or the BBKA Basic

LIFE VICE PRESIDENTS

Mr P O'Reilly
11 Our Lady's Place,
Naas, Co Kildare
045-897568
jackieor@indigo.ie
Mr. M.L. Woulfe
Railway House, Midleton
Co. Cork
021 631011
glenanorehoney@
eircom.net
Mrs Frances Kane
Firmount, Clane,
Co Kildare,
087 2450640
or 045 893150

Certificate must be held for at least one year.
Senior:
Intermediate Certificate and at least five years beekeeping experience.

Intermediate Proficiency Apiary Practical
The Intermediate Proficiency Apiary Practical Examination is intended to be part of a stream that will lead to the Practical Beemasters Certificate. The examination is designed to be less "academic" and there are no written examination papers; (it is not part of the Intermediate Certificate Examination).

The examination will take place in the candidate's own apiary and the Examiners will be two Federation Lecturers appointed by the Executive Council. The pass mark is 70%. 20% of the marks scored may be carried forward to the Practical Beemasters Examination

The prerequisites for Intermediate Proficiency Apiary Practical Examination are: the Preliminary Certificate and at least three years' beekeeping experience satisfactory to the Education Board.

The present prerequisites for the Practical Beemasters Certificate are the Preliminary Certificate and at least five years' beekeeping experience satisfactory to the Examination Board - in the future, an additional prerequisite will be the Intermediate Proficiency Apiary Practical Examination.

Practical Beemaster:
Preliminary Certificate and at least five years' beekeeping experience satisfactory to the Examination Board.

Honey Judge:
Intermediate and Practical Beemaster Certificates, successful showing, having obtained a minimum of 200 points at major shows and a record of stewarding under at least four FIBKA Honey Judges.

Lecturer:
Senior Certificate.

Provincial Examinations
Preliminary and Intermediate examinations will be held at provincial centers on the Saturday closest to 6th April (Intermediate) and May 24th (Preliminary). Please note that the minimum number of candidates for a Centre is **five** for Intermediate and **ten** for Preliminary. Neighbouring associations may pool their candidates to reach those numbers.

A candidate may sit one Intermediate paper at the Provincial Examination and the other paper at the Summer Course.

The fees for all examinations are valid for the year of application only and are listed on the application forms which may be downloaded from the website. In extreme cases, such as illness (a doctor's certificate must be provided); the examination fee may be held over for one year. There are separate entry forms for the http://www.irishbeekeeping.ie/education/proventform.doc Provincial and http://www.irishbeekeeping.ie/education/gormentform.doc Gormanston Summer School Examinations

Fees for Repeat Examinations are the same as for the original examination. Applications to sit the Examinations should be sent to the Education Officer, before the closing dates given above for the Provincial Examinations (applications are however acceptable up to one week after the closing date on payment of a late entry fee which is equal to double the original fee) and before May 1st for the Summer Course Examinations Applications for the Preliminary Examination are also accepted at the Summer Course.

NATIONAL HONEY SHOW
This is held at Gormanston College in conjunction with the annual Beekeeping Course. The Schedule contains 41 Open Classes and 3 Confined classes with €1,000 in prizes. Over 30 Challenge Cups and Trophies are presented for the competition.
Honey Show Secretary: Mr Graham Hall, "Weston", 38 Elton Park, Sandycove, Co Dublin. Tel No (01-2803053) & (087-2406198), E-mail HYPERLINK "mailto:GrahamHall@iolfree.ie" GrahamHall@iolfree.ie

INSURANCE
The limit of indemnity of public liability policy is €6.500, 000 arising from one accident or series of accidents. There is also product liability of €6.500, 000 arising from any one claim. The policy extends to all registered affiliated members whose subscriptions are fully paid up on the 31st December of any one year and whose names are entered in the FIBKA register held by the Treasurer.

ASSOCIATION SECRETARIES

ARMAGH & MONAGH
Mrs. Joanna McGlaughlin
26 Leck Road, Stewartstown
Co Tyrone
BT71 5LS
048-87738702/077-68107984
joanna.mcg@btinternet.com
ATHLONE & DISTRICT
Mr Torquil Sleator
40 Auburn Heights
Athlone
Co Westmeath.
087-2882626
torquil@rd-ireland.com

ASHFORD
Mr Michael Giles
6 The Court, Clonattin Village
Gorey, Co Wexford
086-8369152
michaelgiles46@gmail
BALLYHAUNIS
Mr Phil Bromley
Carrowhawny
Ballyhaunis
County Mayo.
094 9633409
phil@carrowhawny.com

BANNER
Mr Frank Considine
Clohanmore Cree
Kilrush, Co Clare
087-6740462
bannerbees@gmail.com
BEAUFORT
Mr Padruig O'Sullivan
Beaufort Bar & Restaurant
Beaufort, Co Kerry
087-258993006
beaurest@eircom.net

147

FIBKA

✉ ☎

CARBERY
Mr Sean O'Donovan
Drominidy, Drimoleague
Co Cork
087-7715001
seanodonovan10@gmail.com

CARLOW
Mr John Lennon
31 Idrone Park, Tullow Road
Carlow
059-9141315

CO CAVAN
Mr Alan Brady
Shanakiel House,
Drumnagran
Tullyvin, Co Cavan
086-8127920
alan@alanbrady.ie or Info@alanbradyelectrical.com

CO CORK
Mr Robert McCutcheon
Clancoolemore, Bandon,
Co Cork
023-8841714
bob@cocorkbka.org

CO DONEGAL
Mr Dan Thompson
Highfield, Loughnagin
Letterkenny, Co Donegal
074-9125894
dthompson@eircom.net

CO DUBLIN
Mr Liam McGarry
24 Quinn's Road, Shankill
Co. Dublin
087 2643492
mcgarryliam@gmail.com

CO GALWAY
Mr Michael Hughes
Clogher, Claregalway
Co Galway
mhughes@iol.ie

CO KERRY
Mr Ruary Rudd, Westgate
Waterville, Co Kerry
066-9474251
rrudd@eircom.net

CO LIMERICK
Mr Liam Arrigan,
Ballywilliam, Rathkeale
Co. Limerick
086 8070674
liamarrigan@eircom.net

CO LOUTH
Mr Tom Shaw, 201 Ard
Easmuinn
Dundalk, Co Louth
042-9339619/086-2361286
tshaw@iol.ie

CO LONGFORD
Mrs Bridget Koston, Sunnyside
House
Lough Gowna, Co Cavan
043-6683285
dbkoston@eircom.net

CO MAYO
Ms Sue Zajac
Rathnawooran, Killala
Co Mayo
096-34899/087-6260777
mayobeekeepers@mail.com

CO OFFALY
Mr Noel Guerin, Kilbride
Tullamore, Co Offaly
087-2133800
njguerin@gmail.com

CO WATERFORD
Ms Anne Browne
Tower Gate, Tramore
Co Waterford
051-393388/087-6403158
browne_a@eircom.net

CO WEXFORD
Mr Dónal Sammin
Thornhill, Newtown
Enniscorthy, Co Wexford
donalsammin@gmail.com

CHORCA DHUIBHNE
Ms Juli Ni Mhaoileoin
Burnham, Dingle, Co Kerry
086-8337733
julimaloneconnolly@gmail.com

CHONAMARA
Mr Billy Gilmore
Maam West
Leenane, Co. Galway
091-571183/087-7942028
cgilmore@eircom.net

DIGGES & DIST
Ms Mary Hyland
Drumcroman, Drumshanbo
Co. Leitrim
086-2403152
maryhylandhyscapes@eircom.net

DUHALLOW
Mr Andrew Bourke
Pallas, Lombardstown
Mallow, Co Cork
087-2783807
bourke.andy@gmail.com

DUNAMAISE
Mr Derek Banim, Mountain
Farm
Killenure, Mountrath,
Co Laois
086-0856527
derekbanim@yahoo.co.uk

DUNMANWAY
Mr Cormac O'Sullivan
Forrest Oaks, Forrest,
Coachford, Co Cork.
021-7434782/086-4086766
oakforrest@eircom.net

EAST CORK
Mr C Terry
Ait na Greine, Coolbay
Cloyne, Co Cork,
021-4652141
charlesterry@gmail.com

FIBKA

EAST WATERFORD
Mr Michael Hughes
51 Woodlawn Grove
Cork Road, Waterford
051-373461
waterfordbees@gmail.com
FINGAL
Mr John McMullan
34 Ard na Mara Crescent
Malahide, Co Dublin
01-8450193
jmcmullan@eircom.net
FOYLE
Mr P.J. Costello
Lr Drumaiveir, Greencastle
Co Donegal.
074-9381303
pjcost7@eircom.net
GOREY
Ms Cliona Morrish
Coolkenna, Tullow, Co.
Carlow.
086/0874453
cliona.morrish@
goreybeekeepers.com
INISHOWEN
Mr Paddy McDonagh
Milltown
Carndonagh,
Co Donegal.
074-9374881
paddymcdonough@eircom.
net
IVERAGH
Ms Alexis Bowman
The Bungalow
Ballinskelligs, Co Kerry.
0870546481 or 0872745047
lecki@ekit.com
KILLORGLIN
Ms Jane Jackson
Aughils, Castlemaine
Co Kerry
086 104 5979
zanezackson@gmail.com

KILTERNAN
Ms Mary Montaut
4 Mount Pleasant Villas
Bray, Co Wicklow.
01-2860497
yram@connect.ie
MID-KILKENNY
Ms Gypsy Ray
Rhue Lane, Tullabrin
Johnswell, Co Kilkenny
056-775-9935
085-749-1039
gypsyray@eircom.net
NEW ROSS
Mr Seamus Kennedy
Churchtown, Feathard-on
Sea
New Ross, Wexford
051-397259/086- 3204236
seamus.kennedy@yahoo.
co.uk
NORTH CORK
Mr Moss Guiry
Belview, Bruree, Co Limerick
061-397040
mossguiry@yahoo.co.uk
NORTH KILDARE
Mr Patrick Mercer
Ballinagappa, Clane, Co
Kildare
045-868538
pmercer@flogas.ie
NTH TIPPERARY
Mr Jim Ryan
"Innisfail", Kickham Street
Thurles, Co Tipperary
0504-22228
jimbee1@eircom.net
ROUNDWOOD
Mr Paul Jordan
"Rohan", Ballycullen
Ashford, Co. Wicklow
087 1341241
pauljordan1@eircom.net

Sliabh Luachra
Mr Billy O'Rourke
Dooneen, Castleisland, Co
Kerry
066-7141870
siobhancorourke@eircom.
net
SLIGO/LEITRIM
Kieran Mc Donagh
Drumiskabole
Ballisodare, Co Sligo
086-8122534
ktjmcd@hotmail.com
SNEEM
Mr Frank Wallace
Boolananave,
Sneem
County Kerry.
086 3522205
franksneem@hotmail.com
SOUTH DONEGAL
Mr Derek Byrne
Carrick West, Laghey,
Co Donegal.
074-9722340
dcbyrne@eircom.ie
SOUTH KILDARE
Mr Liam Nolan
Newtown, Bagnelstown,
Co Carlow.
059-9727281
liamnolannt@gmail.com
STH KILKENNY
Mr John Langton
Coolrainey
Graiguemanagh,
Co Kilkenny
086-1089652
feganpj@eircom.net
STH TIPPERARY
Mr P J Fegan
Tickinor, Clonmel,
Co Tipperary.
086 1089652
feganpj@eircom.net

149

FIBKA

✉ ☎

STH WEST CORK
Ms Gobnait O'Donovan
38 McCurtain Hill
Clonakilty, Co Cork.
023-8833416/083-3069797
gobnaitodonovan@gmail.com

STH WEXFORD
Mr John Morgan
"An Beacaire", Newhouse
Baldwinstown, Co Wexford
051-563175
morganjbee@eircom.net

SUCK VALLEY
Mr Frank Kenny
Stonepark, Roscommon.
0906-626156.

THE KINGDOM
Ms Elizabeth Ramsden
The Farmhouse, Stradbally
Castlegregory,
Co Kerry
066-7139038
086-1571855
elizram@me.com

THE MIDLAND BKA
Mr Jim Donohoe
11 New Ballinderry
Mullingar,
Co Westmeath
044-9340771/086-2555729
jd@eircom.net

THE ROYAL CO
Mrs Martina Keegan
Grange, Bective, Navan
Co Meath.
046-9029216
martikeegan@eircom.net

WEST CORK
Mr Donal Hanley
Bawnard, Eyeries,
Co Cork
027-74187

WESTPORT
Mr Sean Casey Rosbeg
Westport
Co Mayo
098 26594
absfcasey@eircom.net

150

IBRA ✉ ☎

INTERNATIONAL BEE RESEARCH
ASSOCIATION WEB http://www.ibra.org.uk

I B R A

INTERNATIONAL
BEE RESEARCH
ASSOCIATION

IBRA - International Bee Research Association promotes the value of bees by providing information on bee science and beekeeping. This charity was founded in 1949 and is supported by members from around the world. IBRA owns one of the largest international collections of bee books and journals, as well as the Eva Crane / IBRA historical collection and a photographic collection. It operates an online bookshop, publishes its own books and information leaflets, as well as scientific journals.

CORRESPONDENCE TO:
OPERATIONS DIRECTOR,
Julian Rees
SCIENTIFIC DIRECTOR,
Norman Carreck

16 North Road,
Cardiff,
CF10 3DY
Tel: 029 2037 2409
Fax: 056 0113 5640
Email: mail@ibra.org.uk

PUBLICATIONS
Journal of Apicultural Research
A peer reviewed scientific journal that's worldwide and world class. This quarterly publication contains the latest high quality original research from around the world, covering aspects of biology, ecology, natural history and culture of all types of bees.

Bee World
The flagship publication for IBRA members, this quarterly international journal provides a world view on bees and beekeeping. It covers all topics from bee history to the latest findings in bee science.

IBRA BOOKSHOP
The bookshop is accessible via the web site. To support our charitable status IBRA sells a wide range of publications at competitive prices as well as posters, gifts, DVD's and sundries. IBRA is also a publishing house and offers its members a reduction on IBRA products.

IBRA

✉ ☎

MEMBERSHIP

IBRA is proud of its international status and this is reflected by its members who join from all over the world. The membership package now offers more value than ever before: quarterly issues of Bee World, a discount on IBRA publications and online access to a growing back catalogue. For other benefits and the latest information please visit the web site.

Information about all IBRA publications and services can be found via our web site: www.ibra.org.uk

PLANTS FOR BEES

A Guide to the Plants that Benefit
the Bees of the British Isles

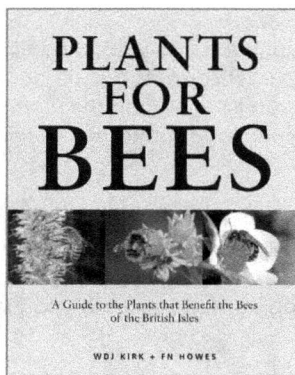

PLANTS
FOR
BEES

A Guide to the Plants that Benefit the Bees
of the British Isles

WDJ KIRK + FN HOWES

W D J Kirk & F N Howes

Foreword by Kate Humble

www.plantsforbees.org

Published and sold by IBRA

INIB

✉ ☎

THE INSTITUTE OF NORTHERN IRELAND BEEKEEPERS (INIB)

www.inibeekeepers.com

Annual Conference and Honey Show. 3rd November 2012
Speakers: Keith Delaplane and Ged Marshall
The Village Centre, Hillsborough BT26 6AR

Objectives of the Institute

The Institute is established to advance the service of apiculture and to promote and foster the education of the people of Northern Ireland and surrounding environs without distinction of age, gender, disability, sexual orientation, nationality, ethnic identity, political or religious opinion, by associating the statutory authorities, community and voluntary organisations and the inhabitants in a common effort to advance education, and in particular:

to raise awareness amongst the beneficiaries about bees, bee-keeping and methods of management;

to foster an atmosphere of mutual support among bee-keepers and to encourage the sharing of information and provision of helpful assistance amongst each other.

Affiliation

INIB is affiliated to the British Beekeepers Association.

With 21,100 members the British Beekeepers Association (BBKA) is the leading organisation representing beekeepers within the UK.

As an INIB member, affiliation gives the following benefits.

- BBKA News
- Public Liability Insurance
- Product Liability Insurance
- Bee Disease Insurance available
- Free Information Leaflets to Download
- Members Password Protected Area and Discussion Forum
- Correspondence Courses
- Examination and Assessment Programme
- Telephone Information
- Research Support
- Legal advice
- Representation and lobbying of Government, EU and official bodies.

Events

The Institute holds an annual conference and honey show. The Institute brings to Northern Ireland world renowned expert speakers from USA and Europe to give talks to beekeepers on the latest research and up to date beekeeping methods.

Education

Demonstrations on various topics such as mead making, preparing honey for shows are held during the year. Courses for honey judges are available.

Honey Bees On Line Studies

INIB has a strong relationship with Professor Jurgen Tautz's of BEEgrouup Biozentrum Universitaet Wuerzburg and his Honey Bee On Line Studies project which continues to develop.

INIB

⊠ ☎

MEMBERSHIP SECRETARY
Lyndon Wortley
Teemore Grange
224 Marlacoo Rd,
Portadown,
BT62 3TD
Membershipsecretary@
inibeekeepers.com

CHAIRMAN
Michael Young MBE
101 Carnreagh,
Hillsborough
BT26 6LJ
02892689724
chairman@
inibeekeepers.com

Holders of the Institute of Northern Ireland Beekeepers Honey Judge Certificate

001. MICHAEL BADGER MBE	01132 945879	BUZZ.BUZZ@NTLWORLD.COM
002. GAIL ORR	02892 638363	GAIL.ORR@BELFASTTRUST.HSCNI.NET
003. CECIL MCMULLAN	02892 638675	MADELINE.MCMULLAN@HOTMAIL.CO.UK
004. HUGH MCBRIDE	02825 640872	LORRAINE.MCBRIDE@CARE4FREE.NET
005. LORRAINE MC BRIDE	02825 640872	LORRAINE.MCBRIDE@CARE4FREE.NET
006. BILLY DOUGLAS	02897 562926	
007. MICHAEL YOUNG MBE	02892 689724	CHAIRMAN@ INIBEEKEEPERS.COM
008. FRANCIS CAPENER	01303 254579	FRANCIS@HONEYSHOW.FREESERVE.CO.UK
009. MARGARET DAVIES	01202 526077	MARG@JDAVIES.FREESERVE.CO.UK
010. IAN CRAIG	01505 322684	IAN'AT'IANCRAIG.WANADOO.CO.UK
011. DINAH SWEET	02920 756483	
013. LESLIE M WEBSTER	01466 771351	LESWEBSTER@MICROGRAM.CO.UK
014. REDMOND WILLIAMS	003535242617	EMWILLIAMS@EIRCOM.NET
015. TERRY ASHLEY	01270 760757	TERRY.ASHLEY@FERA.GSI.GOV.UK
016. IVOR FLATMAN	01924 257089	IVORFLATMAN@SUPANET.COM
017. ALAN WOODWARD	01302 868169	JANET.WOODWARD@VIRGIN.NET
018. DENNIS ATKINSON	01995 602058	DHMATKINSON@TESCO.NET
019 LEO MCGUINNESS	028711 811043	PMCGUINNESS@GLENDERMOTT.COM
020 TOM CANNING	02838 871260	TOM.CANNING@VIRGIN.NET

USA
021. ROBERT BREWER   RBREWER@ARCHES.UGA.EDU
021. BOB COLE

LABRATORY OF APICULTURE & SOCIAL INSECTS (LASI)

UNIVERSITY OF SUSSEX

FURTHER INFORMATION CONTACT
PDr. Francis L. W. Ratnieks,
Professor of Apiculture
Laboratory of Apiculture &
Social Insects (LASI)
Department of Biological &
Environmental Science
University of Sussex, Falmer,
Brighton BN1 9QG, UK

tel: 01273 872954 (landline),
07766270434 (mob)
F.Ratnieks@Sussex.ac.uk

LASI was founded in 1995 and is headed by Dr. Francis Ratnieks, who is the UK's only Professor of Apiculture. Professor Ratnieks received his training in honey bee biology in the USA at Cornell University and at the University of California. Also in the USA, he was a part-time commercial beekeeper with up to 180 hives used for almond pollination and comb honey production.

From 1995 to 2007 LASI was based at the University of Sheffield. In February 2008 Professor Ratnieks moved to the University of Sussex. Sussex University has provided a new laboratory that is ideal for honey bee research. There is a large adjoining apiary with an equipment shed and workshop, and the laboratory is only 100m from the main biology building. There are further apiaries on the university campus just 5 minutes walk away.

LASI is the largest university-based laboratory studying honey bees in the UK and is set up both to do research on honey bee biology and to train the next generation of honey bee scientists. Undergraduate students can do research projects on honey bee biology in their final year, and also receive lectures on honey bee biology. Graduate students can take a PhD in a particular area of honey bee biology. Postdoctoral researchers study honey bees and learn new skills to complement the training they received while doing their PhD.

LASI research focuses on both basic and applied questions in honey bee biology and beekeeping. Research areas include: how honey bees organize their colonies, how they resolve their conflicts, nestmate recognition and guarding, foraging, mating, improved beekeeping techniques, bee health and breeding, conservation of native honey bees.

LASI's mission is to be an international centre of research excellence, to train the next generation of bee researchers, and to be a resource for UK beekeepers and the public in general.

156

(INTER/) NATIONAL PERIODICALS

AMERICAN BEE JOURNAL
(US monthly)
Agents: Northern Bee Books
Scout Bottom Farm,
Mytholmroyd,HX7 5JS
& E.H. Thorne Ltd
Beehive Works Wragby
LN3 5LA
AUSTRALASIAN BEEKEEPER
(Monthly).
Subscriptions US$38
Sample from: Penders PMB
19 Maitland, NSW 2320
Australia
BEE CRAFT
Official monthly journal
of the British Beekeepers
Association
Subscriptions and enquiries to:
Sue Jakeman
Bee Craft Ltd
107 Church St,
Werrington
Peterborough
PE4 6QF
secretary@bee-craft.com
www.bee-craft.com
01733 771221
BEEKEEPERS QUARTERLY, THE
Companion to the
Beekeepers Annual
Subscriptions £26 p.a (but
group schemes at reduced

rates exist for BKAs)
from: Northern Bee Books
Scout Bottom Farm
Mytholmroyd, Hebden
Bridge, W. Yorkshire HX7 5JS
**BERKSHIRE BEEKEEPERS
ASSOCIATIONS, FEDERATION OF
(FBBKA)**
Newsletter Editor,
Mr R F Crocker
25 Shiplake Bottom
Peppard Common
Oxon RG9 5HH
0118 9722315
berksbees@btopenworld.
com
CHESHIRE BEEKEEPER
The Newsletter of CBKA
Mr P Sutcliffe
2 Hatfield Court,
Holmes Chapel
Cheshire CW4 7HP
h.p.sutcliffe@googlemail.
com
GLEANINGS IN BEE CULTURE
US monthly
From: Northern Bee Books
Scout Bottom Farm,
Mytholmroyd,HX7 5JS
& E.H. Thorne Ltd
Beehive Works Wragby
LN3 5LA

**INDIAN BEE JOURNAL IN
ENGLISH**
1325 Sadashiv Peth, Poona
411 8030, India
**INTERNATIONAL BEE RESEARCH
ASSOCIATION**
Journal of Apicultural Research
Journal of ApiProduct and
ApiMedical Science
Bee World
enquires to:
16 North Road
Cardiff
CF10 3DY
mail@ibra.org.uk
IRISH BEEKEEPER
(Monthly) Editor: Jim Ryan
Inisfail, Kickham Street
Thurles, Co. Tipperary
jimbee1@eircom.net
**THE NEW ZEALAND
BEEKEEPER JOURNAL**
Published 11 issues per year
for National BKA of
New Zealand
Contact: Jessica Williams
Executive Secretary
National Beekeepers
Association
PO Box 10792
Wellington 6143
New Zealand

157

NAT MAGS

✉ ☎

P: + 64 4 471 6254
F: + 64 4 499 0876
Tsecretary@nba.org.nz

BEEKEEPER, THE Magazine
of the Scottish BKA.
Membership terms from:
Enid Brown, Milton House,
Main Street, Scotlandwell
Kinross-shire KY13 9JA
Sample copy to view online
www.scottishbeekeepers.
org.uk

SOUTH AFRICAN BEE JOURNAL
Bi-monthly. P.O. Box 41
Modderfontein, 1645, RSA.
THE SPEEDY BEE
Monthly US newspaper,
£24 from: NBB,
Scout Bottom Farm,
Mytholmroyd
Hebden Bridge HX7 5JS

GWENYNWYR CYMRU /
WELSH BEEKEEPER,
EDITOR, Duncan Parks
Cefn Coed
Graianrmon yn lal
Mold CH7 4QW
01824 780504
fax 01824 780822
e-mail,
Duncan@The-Parks.com
The publication of the Welsh
Beekeepers Association
giving news and views of
beekeeping in Wales
and abroad.

Golygydd/Editor: A Duncan
Parks, Cefn Coed, Ffordd
Graianrhyd, Llanarmon yn lal
YR WYDDGRUG CH7 4QW
(01824) 780 504
e-mail,
duncan@the-parks.com
Erthyglau Cymraeg: Dewi
Morris Jones, Llwynderw
Bronant
Aberystwyth SY23 4TG
Manylion tanysgrifau/
Subscription Details:
H. I. Morris, Golygfan
Llangynin, Sancler
CAERFYRDDIN SA33 4JZ
01944 290885

THE NATIONAL DIPLOMA IN BEEKEEPING

The Examinations Board for the National Diploma in Beekeeping was set up in 1954 to meet a need for a beekeeping qualification above the level of the highest certificate awarded by the British, Scottish, Welsh and Ulster Associations.

The Diploma Examination, as designed by the Board, was considered to be an appropriate qualification for a County Beekeeping Lecturer or a specialist appointment requiring a high level of academic and practical ability in beekeeping. It is the highest beekeeping qualification recognised in the British Isles and a high percentage of the past and present holders of the Diploma have given distinguished service to beekeeping education at all levels.

Although the post of County Beekeeping Lecturer has now disappeared, this has merely emphasised the need for some beekeepers to face the challenge of this examination and maintain the high level skills and knowledge needed to keep pace with the increased problems facing all beekeepers at the present time.

The Board consists of representatives from a wide range of organisations and from Government Departments and together form an impressive amalgam of expert knowledge in Beekeeping and Education. Although the National Beekeeping Associations are represented on the Board it is entirely independent of them.

Normally the highest certificate of one of the National Associations is a necessary criterion for eligibility to take the Examination for the Diploma which is held in alternate years. The Written Examination is taken in March, and the Practical, in three sections plus a viva-voce is held in July/August.

The Board also organises an annual Advanced Beekeeping Course covering various parts of the syllabus

HON. SECRETARY
Mrs Margaret Thomas
Tig na Bruaich,
Taybridge Terrace,
Aberfeldy, Perthshire,
PH15 2BS.

CHAIRMAN, Dr David Aston NDB
38 Wressle, Selby
YO8 6ET
01757 638758

NDB

✉ ☎

NDB
National Diploma in | Beekeeping

that are difficult to cover by independent study. Lasting a working week, they cover the main sections of the Syllabus and represent the highest level of training available to British Beekeepers at the present time. The outside lecturers are each acknowledged experts in their particular field. In recent years the Board have been privileged to hold their course at the Fera National Bee Unit at Sand Hutton, York.

For further details regarding the Diploma write, enclosing a stamped A4 SAE to the Secretary, or visit our website: http://www.national-diploma-bees.org.uk/

Those who have gained the National Diploma in Beekeeping

Matthew Allan	Beulah Cullen	G. Howatson	Bill Reynolds
Harry Allen	Celia Davis	Geoff Ingold	Pat Rich
Harrison Ashforth	Ivor Davis	George Jenner	Fred Richards
John Ashton	Alec S.C. Deans	C. F. Jesson	E. Roberts
Dianne Askquith-Ellis	Clive De Bruyn	Simon Jones	Arthur Rolt
David Aston	A.P. Draycott	A.C. Kessel	Jeff Rounce
John Atkinson	M. Feeley	W.E. Large	Graham Royle
Miss E.E. Avey	Barry Fletcher	G.W. Lumsden	J. Ryding
Dan Basterfield	David Frimston	Henry Luxton	J.H. Savage
Ken Basterfield	Oonagh Gabriel	A.S. Mcclymont	Donald Sims
Bridget Beattie	George Gill	J.L. Macgregor	F.G. Smith
Brig. H.T. Bell	Reg Gove	Ian Mclean	George Smith
R.W. Brooke	Eric Greenwood	Ian A. Maxwell	J.H.F. Smith
Norman Carreck	Pam Gregory	Paul Metcalf	Robert Smith
Rosina Clark	Anthony R.W. Griffin	J. Mills	Ken Stevens
Charles Collins	Robert Hammond	Bernhard Mobus	J. Swarbrick
Gerry Collins	Ben Harden	G. Nitonga	Margaret Thomas
Tom Collins	C.A. Harwood	Peter Oldrieve	John Walker
Robert Couston	Leslie Hender	Gillian Partridge	Adrian Waring
John Cowan	Alf Hebden	E.H. Pee	Brian Welch
S. J. Cox	Ted Hooper	L.E. Perera	J. Wilbraham
Jim Crundwell	Geoff Hopkinson	E.R. Poole	

NHS

THE NATIONAL HONEY SHOW

www.honeyshow.co.uk
THE 2013 SHOW IS AT ST GEORGE'S COLLEGE,
WEYBRIDGE, SURREY KT15 2QS
24TH – 26TH OCTOBER 2013.

This venue is excellent

Just off the M25 junction 11
Rail from Waterloo to Weybridge or Addlestone

Free car parking

The Show itself is a wonderful competitive exhibition
of all the products of the bee-hive, coupled with an
excellent series of lectures, workshops and a wide
variety of trade and educational stands.

We recommend that you attend all three days, and
suggest that you become a member of the Show –
just **£12.00** per annum

For further information, please write to the Hon General
Secretary, or Email: showsec@zbee.com or visit our
website www.honeyshow.co.uk

HON. GENERAL SECRETARY
REV. H.F CAPENER
1 Baldric Road
Folkestone CT20 2NR

HON TREASURER
C S Mence
27 Acacia Grove
New Malden, Surrey KT3 3BJ

PUT THIS DATE IN YOUR DIARY
27TH – 29TH OCTOBER 2012

THE.NATIONAL
HONEY.SHOW

www.honeyshow.co.uk
THE 2013 SHOW IS
AT ST GEORGE'S COLLEGE,
WEYBRIDGE
24TH - 26TH OCTOBER

RR

✉ ☎

ROTHAMSTED RESEARCH
www.rothamsted.ac.uk

ROTHAMSTED
RESEARCH

ROTHAMSTED RESEARCH
Department of AgroEcology,
Rothamsted Research
Harpenden,
Hertfordshire.
AL5 2JQ

STAFF
Dr Juliet Osborne*
Dr Alison Haughton
Dr Peter Kennedy**
Dr Mathias Becher***
Dr Stephan Wolf
Dr Samantha Cook
Jenny Swain
Emma Wright+
Jonathan Carruthers (PhD student)
Steve Kennedy
Sue Bird

* relocated to the Environment & Sustainability Institute [ESI] at the University of Exeter at Penryn, Cornwall
** at ESI from February 2013
+ until Nov 2012

Rothamsted is the oldest laboratory in the world devoted to agricultural research, having been established in 1843. Research on bees has been continuous since 1923 and current expertise is founded on pioneering work at Rothamsted by a number of eminent bee scientists. Approximately 20 honey bee colonies are now maintained for experimental purposes and the Tomkins Field Laboratory houses a behaviour laboratory, observation hive room and bee flight room as well as a workshop for the manufacture of specialised equipment.

The Rothamsted site provides a unique working environment with specialist modern equipment facilitating research on plant and microbial metabolites, molecular biology and synthetic and analytical chemistry. There is an experimental farm for complex field experiments, and there is a suite of glasshouses, controlled environment facilities, an insectary and a state of the art bioimaging suite housing three new electron microscopes and a confocal laser scanning microscope. Experimental design and analysis are backed up by excellent statistical, computing and library support.

BEE BEHAVIOUR AND POLLINATION ECOLOGY

We are investigating the interaction between bees, crops and the agricultural environment. The spatial and temporal foraging behaviour of honey bees and bumble bees within agricultural areas is being compared. Harmonic radar is being used to track flying bees, and other pollinators such as butterflies, to obtain new information about their flight paths, forage ranges, food preferences and orientation mechanisms.

An integrated model for predicting bumblebee population success and pollination services in agro-ecosystems will be developed by the Environment and Sustainability Institute (ESI) at the University of Exeter, Falmouth campus, Rothamsted and the University of Stirling, and will provide a powerful tool for shaping recommendations for land managers and policy makers for the sustainable spatial management of pollination within arable and horticultural production systems.

Various qualities of different varieties of crops (oilseed rape and short rotation coppice willows) as important resources for bees are being investigated. The nutritional value of the nectars and pollens, effects on bee fitness and behaviour are key areas of interest.

HONEY BEE PATHOLOGY
Rothamsted's research on the natural history and epidemiology of the infections and parasites of bees has had wide international recognition. However, research on honey bee pathology is currently suspended due to changes in funding available from Defra for bee health. Over the last 20 years, this work focused on *Varroa destructor* and the losses caused by honey bee virus infections that the mite transmits. In a collaborative project with Horticulture Research International (at University of Warwick), investigating potential biological control agents of *V. destructor*, the research identified and characterised fungal pathogens which are active against the mite but which are relatively safe for bees and other beneficial insects. Biological control offers an environmentally acceptable approach to the problem that could have considerable economic benefits, and we are actively seeking funding to continue this work.

A studentship has been examining the interaction between disease, individual honey bee behaviour and behaviour or survival of the colony. A recently awarded Insect Pollinator Initiative project is assessing the impact

RR

✉ ☎

of emergent diseases, including the *Varroa* associated Deformed wing virus, and the Microsporidian *Nosema ceranae* on the flight performance and orientation ability of honeybees and bumblebees and its consequences for bee populations.

INFORMATION EXCHANGE

Expertise in bee research is drawn upon by scientific colleagues world-wide and there are research links with institutes and universities in this country and abroad. Research findings are published in scientific journals but popular articles are also written for the beekeeping and agricultural press. Staff members serve on both national and international committees on diverse aspects of apiculture and a vigorous programme of lectures presented to national and local beekeeping associations ensures effective communication of recent work.

FUNDING

Rothamsted receives funds for research from the Biotechnology and Biological Sciences Research Council, through competitions and contracts from the Department for Environment, Food and Rural Affairs, the European Community, from Levy boards, commercial and other organisations. The support of the bee research programme in recent years by grants from the British Beekeepers Association, C. B. Dennis British Beekeepers Research Trust, the Eastern Association of Beekeepers and the Bedfordshire, Cambridgeshire, Norfolk, St Albans and Hertfordshire Beekeepers Associations is gratefully acknowledged.

For more information visit: **http://www.rothamsted.ac.uk**

THE SCOTTISH BEEKEEPERS' ASSOCIATION

AIMS OF THE ASSOCIATION
- publish a monthly magazine
- maintain the Moir Library in Edinburgh
- conduct examinations in the art of beekeeping
- provide insurance and a compensation scheme for members

EDUCATION
The SBA arranges courses and awards certificates to successful candidates in the Scottish Basic Beemaster, Expert Beemaster, Honey Judge and Microscopy Examinations. It also actively promotes beekeeping by informing the public, especially the young, about bees and their benefits to the environment.

INSURANCE AND THE COMPENSATION SCHEME
All members of the SBA have insurance against Public Liability. The SBA Compensation Scheme is restricted to bee colonies located in Scotland and allocates part-replacement value for damage by vandalism, fire, theft and certain brood diseases.

LIBRARY
The SBA Moir Library in Edinburgh has one of the world's finest collection of beekeeping books. A library card is issued annually to every member who can borrow books at the cost of return postage only. Details may be obtained from the Library Convener.

MARKETS
Advice is given on all aspects of marketing honey products at appropriate times. Suggested bulk, wholesale and retail prices are notified in the magazine.

GENERAL SECRETARY
Mrs. Bronwen Wright
20 Lennox Road
Edinburgh EH5 3JW
0131 552 3439
secretary@
scottishbeekeepers.org.uk

HON PRESIDENT
The Rt. Hon. Earl of Mansfield D.L, J.P
Scone Palace
Perth PH2 6BE

HON. VICE PRES,
Iain F Steven
4 Craigie View
Perth
PH2 0DP
01738 621100
Ian Craig
30 Burnside Ave, Brookfield,
Johnstone, Renfrewshire,
PA5 8UT
01505 322684
beekeeper30@btinternet.
com

HON. LIBRARIAN
Mrs. Margaret M. Sharp
City Librarian, City Library
George IV Bridge, Edinburgh

HON. LEGAL ADVISER,
Taggert, Meil & Mathers
20 Bon Accord Sq,
Aberdeen
01224 588020

165

SBA

INDEPENDENT EXAMINER
Lynne Ramsay
Lynne Ramsay,
11, Chancelot Terrace,
Edinburgh EH6 4SS
0131 552 8218

PUBLICATIONS
- The Scottish Beekeeper is published monthly and sent post free as part of the annual membership fee of £30 payable to the Membership Convener.
- Introduction to Bees and Beekeeping is £6.00 plus postage and may be obtained from the Advertising and Publicity Convener.

PUBLICITY
Members can purchase the Association tie, lapel badge, car sticker etc. Details may be obtained from the Advertising and Publicity Convener.

SHOWS
Three major annual honey shows are held in Scotland. They are at the Royal Highland Show, Ingliston, Edinburgh in June, while the Scottish National Honey Show and the East of Scotland Honey Show are both held at the Dundee Flower and Food Festival in September. Shows are also held at Aberdeen, Ayr, Inverness and there are 2 shows in Fife.

Executive Committee

PRESIDENT,
Phil McAnespie
12 Monument Road
Ayr KA7 2RL
01292 885660
membership@
scottishbeekeepers.org.uk

VICE PRESIDENT,
Mrs. Bronwen Wright
20 Lennox Road,
Edinburgh EH5 3JW
0131 552 3439
secretary@
scottishbeekeepers.org.uk

IMM. PAST PRES,
Ian Craig
30 Burnside Avenue
Brookfield, Johnstone
Renfrewshire PA5 8UT
01505 322684
beekeeper30@
btinternet.com

GENERAL SEC
Mrs. Bronwen Wright
20 Lennox Road, Edinburgh
EH5 3JW
0131 552 3439
secretary@
scottishbeekeepers.org.uk

SBA CO-ORDINATOR,
Iain F. Steven
4 Craigie View, Perth PH2 0DP
01738 621100
lomand@btinternet.com

TREASURER, RACHAEL GIBBINS,
21 Gilmerton Dykes
Gardens, Edinburgh
EH17 8JL
0131 664 5658
sbatreasurer@hotmail.co.uk

EDITOR, SCOTTISH BEEKEEPER,
Nigel Southworth
47 Middleton Road, Uphall,
Edinburgh, EH52 5DF
01506 865762
editorscottishbeekeeper@
gmail.com

CONVENERS OF STANDING COMMITTEES

MEMBERSHIP CONVENER
P. McAnespie
12 Monument Rd.Ayr
KA7 2RL   01292 885660
membership@
scottishbeekeepers.org.uk

INSURANCE & COMPENSATION
C. Irwin
55 Lindsaybeg Rd
Chryston, Glasgow
G69 9DW
0141 7791333
ceirwin@talktalk.net

ADVERTISING & PUBLICITY
Miss E Brown
Milton House, Main Street
Scotlandwell, Kinross
KY13 9JA   01592 840582
honeybees@onetel.com

EDUCATION, Alan Riach
Woodgate, 7 Newlands
Ave, Bathgate
EH48 1EE
01506 653839
alan.riach@which.net

PROMOTION OF BEEKEEPING
CONVENER,
P Mathews
4 Annanhill
Annan
Dumfries-shire
DG12 6TN
01461 205525
silverhive@hotmail.com

SHOWS, Miss E Brown
Milton House, Main Street
Scotlandwell, Kinross
KY13 9JA
01592 840582
honeybees@onetel.com

LIBRARY, Mrs Una Robertson
13 Wardie Ave
Edinburgh
EH5 2AB
una.robertson@btinternet.
com

MARKETS, John Durkacz
15 Lundin Road
Crossford
Fife KY12 8PW
01383 722186
Durkacz@hotmail.co.uk

BEE HEALTH, Phil Moss
Ealachan Bhana
Clachan Seil
Oban
PA34 4TL
01852 300383
phil.moss@dsl.pipex.com

ICT CONVENER,
Alasdair Joyce
Manachie Lodge.
Dallas Dhu
Forres
IV36 0RR
01309 671288
webmaster@
scottishbeekeepers.org.uk

AREA REPRESENTATIVES
NORTH,
Mrs Sheila Barnard
Viewmount, Tobermory,
Isle of Mull
PA75 6PG
01688 302008
tim-barnard@lineone.net

EAST, John Trout
13 Middlebank Holdings
Dunfirmline, Fife
KY11 8QN
01383 415534
jtout@middlebank.
demon.co.uk

WEST, Mike Thornley
Glenarn House, Glenarn
Road, Rhu, Helensburgh
G84 8LL
01436 820493
masthome@dsl.pipex.com

ABERDEEN AND MORAYSHIR,
Dr Stephen Palmer
Fintry School House,
Fintry, near Turiff
AB53 5RN
01888 551367
palmers@fintry.plus.com

SBA

✉ ☎

S.B.A LECTURERS ★Addresses in SBA Honey Judges List

All those listed claim expenses (except G. Sharpe, Bees adviser funded by SGRPID),
All speakers accompany talks with visual aids

★ MISS. E. BROWN (General)
01592 840542
★ M BADGER (General)
0113 2945879
★ I. CRAIG (General)
01505 322684
A.B. FERGUSON
(General, Varroa)
Firparkneuk. Kirtlebridge
Lockerbie   DG11 3LZ
01461 500322
★ C. IRWIN (General)
0141 7791333
★ DR. F. ISLES (Bee diseases)
01382 370 315
M.M. PETERSON
(Bee genetics)
Balhaldie House,
High street, Dunblane
FK15 0ER
01786 822093

MRS. U. A. ROBERTSON
(History of SBA, Moir Library,
History of beekeeping)
13 Wardie Ave
Edinburgh EH5 2AB
0131 552 5341
G. SHARPE (SAC) (Varroa
Management: My apiary
management system)
Apiculture Specialist
Life Science Technology
Group, SAC Auchincruive
Ayr KA6 5HW
01292 525375
Mrs M Thomas (General)
Tighnabraich, Taybridge
Terrace, Aberfeldy
Perthshire PH15 2BS
01887 829710

J. TYLER (Strain selection
and queen breeding)
22 Montgomerie Drive
Fairlie, Ayrshire
01475 568421
★ L. M. WEBSTER (General)
01466 771351
DR G RAMSAY (Beekeeping on
the Internet / Can Bees fight
Varroa?)
Parkview, Station Road
Errol, Perth   PH2 7SN
01821 642385
A RIACH
(Beehives through the Ages)
Woodgate, 7 Newland Ave
Bathgate
EH48 1EE
01506 653839

MEMBER ASSOCIATIONS AND THEIR SECRETARIES

ABERDEEN, Rosie Crighton
29 Marcus Cresc
Blackburn, Aberdeen
AB21 0SZ
01224 791181
rosie@crighton-finflater.
fsbusuness.co.uk
AYR, Mrs L Baillie
Windyhill Cottage
Uplands Rd, Sundrum
Ayre, KA6 5JU
01292 570659
lbaillie@sundrum.demon.
co.uk
BORDER, Norman Jarvis
6 Dean Road, Sprouston,
Kelso TD5 8HN
01573 228276
noel.tweedview@
btopenworld.com

BUTE, Alison Cross
Marionslea, Minister's Bray,
Rothsay, Isle of Bute
PA20 9BG
01700 504627
alison.cross2@virgin.net
CADDONFOOT,
James & Julia Edey
West Water, Bedrule, Hawick,
Roxburghshire,
TD9 8TD
01450 870400
jamesedey@googlemail.com
CLYDE AREA, Mr George Morrison
102 Woodside Ave Bearsden
G61 2NZ
0141 942 9419

COWAL, Brian Madden
123a Alexandra Parade
Dunoon, PA23 8AW
01369 703317
brian_maden@btinternet.com
DINGWALL, Alpin Stewart
Rowan Cottage,
Fasaig, Torridon by
Achnasheen, Ross-shire
IV22 2EZ
01445 791450
dingwall.beekeeping@
googlemail.com
DUNBLANE & STIRLING,
Fiona Fernie
Greystones Dunira,
By Comtie
PH6 2JZ
01764 679152
secretary@dsbka.net

DUNFERMLINE & WEST FIFE
Dr T Scott
Grange Farmhouse
Grange Rd, Dunfermline
KY11 3DG
01383 733125
dwf@fifebeekeepers.co.uk
EAST OF SCOTLAND
Mrs Susan Anderson
12 Kirkgate, Letham,
Forfar DD8 2XQ
01307 819477
anderson44 @talktalk.net
EAST LOTHIAN, Mrs Jo Dodds
20 Kings Avenue
Longniddry, East Lothian
EH32 0QN
01875 852916
eastlothianbeekeepers@gmail.
com
EASTER ROSS
Mrs P Douglas-Menzies
Cardboll Cottage, Fearn
Ross-shire, IV20 1XP
01862 871572
pruedm@gmail.com
EASTWOOD, Jeanette Malcolm
157 Campbell Avenue,
Glasgow
G41 3DR
0141 632 3133
jean@hotmail.co.uk
EDINBURGH & MIDLOTHIAN
P Steven
Eastercowden Cottage
Dalkeith, Midlothian
EH22 2NS
07703 528801
porrsteven@yahoo.co.uk
FIFE, Janice Furness
The Dirdale, Boarhills
St. Andrews, Fife KY16 8PP
01334 880 469
jcfurness@dirdale.fsnet.co.uk

FORTINGALL, Mrs. Jo Pendleton,
Lilac Cottage
Old Bridge of Tilt by Pitlochry
PH18 5TP
01796 481 362
d.h.pendleton@btinternet.com
GLASGOW DISTRICT,
Mr P Stromberg
21 Woodside Houston,
Renfrewshire PA6 7DD
01505 613830
pstromberg1@aol.com
HELENSBURGH, Gordon Smith
The Moorings, Ferry Road,
Rhu G84
07980 578206
gordon@windsmiths.co.uk
INVERNESS-SHIRE
Stella Chisholm
Gladstone Cottage,
Struy,Beauly IV4 7JU
01463 761251
ibasecretary@hotmail.co.uk
KELVIN VALLEY, I Ferguson
4 South Glassford Street
Milngavie G62 6AT
0141 956 3963
jeanian@ferguson2007.plus.
com
KILBARCHAN AND DISTRICT
I. Craig
30 Burnside Ave
Brookfield
Johnstone PA5 8UT
01505 322684
beekeeper30@btinternet.com
KILMARNOCK & DISTRICT
J. Campbell
North Kilbryde House
Stewarton
Kilmarnock KA3 3EP
01560 482489
john.d.campbell@talktalk.net

KIRRIEMUIR,
'Disbanded'
LARGS & DIST, Kate Dahlstrom
3 Burnside Road
Largs KA30 9BY
01475 740437
k.dahlstrom@btinternet.com
LAMANCHA, Joyce Jack,
23 South Park,
West Peebles, EH45
01721 722444
joycecjack@aol.com
LOCHABER, Rev Kate Atchley
Anasmara, Mingarry
Acharacle, Inverness-shire
PH36 4JX
01967 431420
contact@kateatchley.co.uk
MORAY, T Harris
Cowiemuir, Fochabers
01343 821 282
tonyharris316@btinternet.com
MULL, Mrs. S. Barnard
Viewmount, Tobermory
Isle of Mull PA75 6PG
01688 302008
tim-barnard@lineone.net
NAIRN & DISTRICT, John Burns
Woodlands, Cawdor Road
Nairn IV12 5EF
01667 454887
jandjburns@hotmail.com
OBAN & DISTRICT,
Phil Moss
Ealachan Bhana
Clachan Seil
Oban PA34 4TL
01852 300383
phil.moss@dsl.pipex.com
OLRIG and District, Robin Inglis
Roadside Skirza
Freswick, Wick KW1 4XX
01955 611260
gailinglis@btinternet.com

SBA

✉ ☎

ORKNEY, Doris Fischler
84 Victoria street
Stromness
Orkney KW16 3BS
01856 850447
d@orcahotel.com

PEEBLES-SHIRE,
Amanda Clydesdale
20 Kingsmeadows Gardens
Peebles EH45 9LB
01721 720563
amanda.clydesdale@
btinternet.com

PERTH AND DISTRICT ,
Linda Legget
2 Acharn, Perth
PH1 2SR
01738 580024
caley.blue@sky.com

SKYE & LOCHALSH, M Purrett
15 Glasnakille, Egol
Isle of Skye, IV49 9BQ
01471 866 207
jenny@scotnet.co.uk

S. OF SCOTLAND, Edith Reyntiens
40 George Street, Dumfries
DG1 1EH
01387 266583
edith@georgestdesign.co.uk
fergiearchie@tiscali.co.uk

SUTHERLAND, Sue Steven
Mulberry Croft, 2 East Newport,
Berriedale Caithness KW7 6HA
01539 751 245
mulberrycroft607@
btinternet.com

WEST'N GALLOWAY,
Linda Robertson
Craigenveoch Farm, Glenluce,
Newton Stewart, DG8 0LD
07825 51 4 726
lindaglenmheran@aol.com

WEST LINTON & DISTRICT
D. Stokes
100 Main Street, Roslin
Midlothian EH25 9LT
0131 440 3477
wlbka@live.co.uk

WESTERN ISLES, Elizabeth Shelby,
6 Earshader,
Isle of Lewis, HS2 9LP
01851 612 239
Elizabeth:bshelby@
demon.co.uk

Freuchie BKA disbanded

SBA ACTIVE HONEY JUDGES

M BADGER
Kara, 14 Thorn Lane,
Roundhay,
Leeds LS8 1NN

MISS E. BROWN
Milton House, Main Street,
Scotlandwell
Kinross KY13 9JA
01592 840582

P.J. BROWNE
The Rowan Tree, Gairlochy
Spean Bridge
Inverness-shire PH34 4EQ
01397 712730

M. CANHAM
Whinhill Farm House
by Cawdor, Nairn IV12 5RF
01667 404314

I. CRAIG
30 Burnside Avenue
Brookfield, Johnstone
Renfrewshire,PA5 8UT
01505 322684

H DONOHOE
7 Grant Road
Banchory AB31 5UW
01330 823502

C. E. IRWIN
55 Lindsaybeg Road
Chryston, Glasgow
G69 9DW
0141 7791333

DR F. ISLES
"Gardenhurst",
Newbigging Broughty Ferry
Dundee DD5 3RH
01382 370315

P MATHEWS
MRS C MATHEWS
4 Annanhill
Annan, Dumfries-shire
DG12 6TN
01461 205525

MS B L MCLEAN
Upper Flat, 2 Invererne Rd,
Forres IV36 1DZ
01309 676316

W.B. TAYLOR
West Newbigging Cottage,
Glenbervie Road, Drumlithie
Stonehaven AB39 3YA
01569 740375

L.M. WEBSTER
Birchlea, Rothiemay, Huntly
Aberdeenshire AB54 5LN
01466 771351

C. WEIGHTMAN
Shilford, Stocksfield,
Northumberland NE43
4HW
01661 842082

C. WILSON
Cedarhill, Auchencloch,
Banknock, Bonnybridge
FK4 1VA
01324 840227

DR D WRIGHT
MRS B WRIGHT
20 Lennox Row
Edinburgh EH3 5JW
0131 552 3439

M. YOUNG
101 Carnreagh,
Hillsborough
County Down
N. Ireland BT26 6LJ
0289 268972

UBKA
✉ ☎

ULSTER BEEKEEPERS' ASSOCIATION

www.ubka.org

OBJECTS OF THE ASSOCIATION

The objects of the Association are to unite beekeepers for their mutual benefit to serve the best interests of beekeeping by all means within its power and to foster its healthy development.
For the purpose of achieving these objects the Association will:

- promote the formation of local Beekeepers' Associations
- disseminate information and advice about beekeeping
- provide examination facilities in the craft of beekeeping
- encourage maintenance and improvement of the beekeeping environment.

EDUCATION

In conjunction with the College of Agriculture, Food & Enterprise (CAFRE), the U.B.K.A. assists in organising classes for Preliminary, Intermediate and Senior Certificate Examinations in Beekeeping following the syllabus of the Federation of Irish Beekeepers' Associations (FIBKA).

INSURANCE

Affiliated local Associations and their individual members have access to the UBKA group public and product liability insurance scheme.

APIARY SITES

Almost all twelve local Associations and CAFRE's Greenmount Campus have access to apiary sites and, for some sites, access to observation houses provided with help from Leader 2 funding, for use in demonstrating and promoting good practice to members, schools and other interested groups.

PRESIDENT,
David Wright
24 Quarry Road
Lisbane, Comber,
Newtownards
Co Down BT23 5NF
CHAIRMAN, John Witchell
40 Hollywood Road,
Newtownards
Co. Down BT23 4TQ
TREASURER, Matthew Porter.
375 Old Glenarm Road
LARNE Co Antrim
BT40 2LH
LECTURERS
Jim Fletcher
26 Coach Road, Comber
Co.Down. BT23 5QX
Ethel Irvine
2 Laragh Lee
Ballycassidy
ENNISKILLEN
BT94 2JT
Lorraine McBride
11, Ballyloughan Park
Ballymena, Co.Antrim,
BT43 5HW
Norman Walsh
43, Edentrillick Rd
Hillsborough, Co. Down
BT26 6PG

171

UBKA

✉ ☎

LECTURERS CONTINUED
Rev Sam Millar
41 Rectory Park
Garvagh, COLERAINE
Co Londonderry
BT51 5AJ

HONEY JUDGES
Jim Fletcher
26 Coach Road,
Comber
BT23 5QX
Michael Young
Mileaway, Carnreagh Road
Hillsborough, Co. Down
BT26 6LJ
Norman Walsh
43 Edentrillick Rd
Hillsborough Co. Down
BT26 6NH

HONEY SHOWS

Local Associations stage honey shows throughout Northern Ireland. The Northern Ireland Honey Show, hosted by the Belfast City Parks Department, is held annually in September in the Botanic Gardens Belfast.

CONFERENCE

The 69th UBKA Annual Conference will be held on 8 – 9th March 2013 at CAFRE's Greenmount Campus, Antrim. Contact the U.B.K.A. Conference Manager at 07871 161303 and www.ubka.org for details.

SECRETARIES OF ASSOCIATIONS

Belfast,
Alan Rea
12 Kirkliston Drive
Belfast BT5 5NX

Clogher Valley, Trevor Watson
14 Findrum Road,
Ballygawley, Dungannon,
Co. Tyrone BT70 2JL

Derry City, Billy Chambers
23 Hatmore Park,
Derry, BT48 0AY.

Dromore, Patrick Lundy
116 Dromore Road,
Ballynahinch,
Co.Down BT24 8HK

East Antrim, Desmond Blair
21 Ballylesson Road,
Larne, Co Antrim

Fermanagh, Brian Richardson
Agho,305 Lattone Road,
Belcoo, Enniskillen
Co. Fermanagh

Killinchy, Susanna Best
2 Dufferin Villas,
Bangor BT20 5PH

Mid Antrim, Angela Morrow
23 Beechwood Drive Ahogill
Ballymena BT42 1NB

Mid Ulster, Anne Milligan
61 Blackisland Road ,
Annaghmore,
Portadown BT62 1NE

Randalstown, Susie Hill
7 Nutts Corner Road,
Crumlin,
Antrim BT29 4BW

Roe Valley, Sandra Logan
22 Knocknougher Road,
Macosquin, Coleraine,
Co. Londonderry.

Rostrevor & Warrenpoint,
Christina Joyce "The Grange",
1 Mourne Park, Kilkeel,
Co Down, BT34 4LB

WBKA/CGC
✉ ☎

CYMDEITHAS GWENYNWYR
CYMRU WELSH BEEKEEPERS' ASSOCIATION

AMCANION Y GYMDEITHAS / AIMS OF THE ASSOCIATION
- Promote and develop beekeeping in Wales
- Conduct examinations in beekeeping
- Liaise with organisations and bodies for the benefit of beekeeping in Wales

AELODAETH UNIGOL / INDIVIDUAL MEMBERSHIP
Individual membership of the WBKA is provided for persons who do not live within the areas of branch associations, and wish to support the association. Information relating to benefits and facilities provided for individual members is available from the Individual Membership Secretary.

ARHOLIADAU / EXAMINATIONS
The Examinations Board conducts six grades of examinations: Junior, Primary, Intermediate, Practical, Honey Show Judges, Senior. Information is available from the Examination Board Secretary.

Candidates following the Duke of Edinburgh Award Scheme may receive information regarding the inclusion of beekeeping as a course submission from the Examinations Secretary.

CYNHADLEDD/ CONVENTION
At the Royal Welsh Agricultural Society's Showground, Llanelwedd. This event is normally held during Late March/ Early April. Information relating to this event is available from the convention secretary.

YSWIRIANT / INSURANCE
All individual and fully paid up members of beekeeping associations affiliated to WBKA are covered against 'Public and Product' liability claims. All affiliated associations are covered against public liability during conventions officially organised by the association.

YSGRIFENNYDD / SECRETARY
John Page
The Old Tannery
Pontsian
Llandysul
Ceredigion
SA44 4UD
secretary@WBKA.com

LLYWYDD/PRESIDENT
Dinah Sweet,
Graig Fawr Lodge
Caerphilly
CF83 1NF
president@wbka.com

CADEIRYDD/CHAIR
Valerie Forsyth
Bwlch y Rhyd
Nanternis
New Quay
SA45 9RS
chair@wbka.com

IS-GADAIRYDD/
VICE CHAIR
Tom Pegg
depchair@wbka.com

WBKA/CGC

✉ ☎

TRYSORYDD/TREASURER
Vincent Frostick

GWEFEISTR/WEBMASTER
AND GOLYGYDD/EDITOR
Emmanuel & Rebecca Blaevoet
editor@wbka.com

IS-OLYGYDD (ERTHYGLAU
CYMRAEG)/SUB EDITOR
Dewi Morris Jones
Llwynderw, Bronant
Aberystwyth SY23 4TG
(01974 251264)

ARHOLIADAU/EXAMINATIONS
Dinah Sweet
Graig Fawr Lodge
Caerphilly
CF83 1NF
president@wbka.com

The WBKA Individual Membership benefits include cover under the BDI Scheme against the loss, due to foul brood diseases, of a minimum number of stocks (determined by BDI). Affiliated Associations provide this cover for their members.

LLYFRGELL / LIBRARY
The reference sections of all county libraries in Wales have details of the names and addresses of Secretaries of Associations affiliated to WBKA.

Books on beekeeping can be borrowed from county, branch and mobile libraries. The Library, Ffordd y Bala, Dolgellau LL40 2YS, has been nominated to stock beekeeping books.

Members of associations affiliated to IBRA may borrow books/documents from its library.

GWENYNWYR CYMRU - The Welsh Beekeeper
A publication of the Welsh Beekeepers Association, giving news and views of beekeeping and related subjects. Articles and advertisements enquiries should be sent to the Editor. Articles written in Welsh should be sent to the Sub Editor. Gwenynwyr Cymru is provided free to members of Affiliated Associations and Individual Members. Information regarding subscriptions is available from the Individual Membership / Subscription Secretary.

GWASANAETH CLYWELED / AUDIO-VISUAL AIDS SERVICE
This service is available to all affiliated associations and individual members. Further information is available from the Audio-Visual Aids Secretary.

DARLITHWYR / DANGOSWYR, LECTURERS / DEMONSTRATORS
The names and addresses of lecturers and demonstrators, recommended by associations affiliated to the WBKA, are available from the General Secretary.

CYNLLUN CYSWLLT CHWYSTRELLU / SPRAY LIAISON SCHEME
Information is available from the General secretary

174

WBKA/CGC

✉ ☎

SIOEAU / SHOWS

Honey/beekeeping sections are included at the Royal Welsh Agricultural Show, Llanelwedd, (OS ref: SO040520) during July, and at county, town and village shows throughout Wales. Information relating to these events may be obtained from secretaries of associations in the locality of the shows.

The historic FFAIR FEL ABERCONWY is held annually in the main street of the town, (OS ref: SH278378), on 13th September. Further information is available from the secretary of Conwy Association.

RHEOLAU CYFREITHIOL / STATUTORY REGULATIONS

The administration of the statutory regulations governing all aspects of beekeeping in Wales, is the responsibility of the Wales National Assembly, Caerdydd, CF99 1NA Phone (02920) 825111 Fax: (02920) 823352 Matters concerning statutory regulations, their implications and execution, should be addressed to the Minister of Agriculture and Rural Affairs, Wales National Assembly, at the above address.

THE.NATIONAL

HONEY.SHOW

www.honeyshow.co.uk
THE 2013 SHOW IS
AT ST GEORGE'S COLLEGE,
W E Y B R I D G E
24TH - 26TH OCTOBER

AELODAETH UNIGOL - TANYSGRIFAU / INDIVIDUAL MEMBERSHIP SUBSCRIPTIONS
Ian Hubbuck
White Cottage
Berriew
SY21 8BB
01686 640205
ianhubbuck@hotmail.com

INSURANCE:
Rhodri Powell
146 Pandy Rd, Bedwas,
Caerphilly CF83 8EP
rhodri.gp.powell@sky.com

AUDIO VISUAL AIDS:
F. G. Eckton
Cartref
Llanafan Fawr,
Llanfair ym Muallt LD2 3LT
01591 620456

CONVENTION SECRETARY:
Sanna Burns
sanna.orlando@virgin.net

CONVENTION TRADE STANDS SECRETARY:
Wally Shaw
Llwyn Ysgaw, Dwyran,
Llanfairpwll, Anglesey
LL61 6RH 01248 430811
waltershaw301@btinternet.com

WBKA/CGC

✉ ☎

CYMDEITHASAU TADOGOL A'U YSGRIFENYDDION / AFFILIATED ASSOCIATIONS AND SECRETARIES

ABERYSTWYTH, Ann Ovens,
Tan-y-Cae, Nr Talybont,
Ceredigion,
SY24 5DP
01970 832359
ann.ovens@btinternet.com
ANGLESEY, Ian Gibbs
Dryll, Bodorgan
Ynys Mon
LL62 5AD
01407 840314
secretaryabka@gmail.com
BRECKNOCK AND RADNOR,
Dr Gillian Todd, Meadow
Breeze, Llanddew, Brecon
LD3 9ST
01874610902 07971314798
gbtodd@btinternet.com
BRIDGEND, Sue Verran Ty Mel,
Maesteg Rd. Bridgend
CF32 0EE
01656 729699
verran@btinternet.com
CARDIFF AND VALE, Annie
Newsam
Stonecroft, Mountain Road,
Bedwas, Caerphilly, CF83 8ER
annienewsam@hotmail.co.uk
CARMARTHEN, Brian Jones
Cwmburry Honey Farm,
Ferryside, Carmarthenshire,
SA17 5TW
01267 267318
beegeejay2003@yahoo.co.uk

CONWY, Mr Peter McFadden,
Ynys Goch
Ty'n y Groes,
Conwy LL32 8UH
01492 650851
peter@honeyfair.freeserve.
co.uk
EAST CARMARTHEN
Linda Wallis,
Maestroyddyn Fach
Harford
Llanwrda
Carmarthenshire
SA19 8DU
01558 650774
linandbaz@aol.com
FLINT AND DISTRICT,
Jill and Graham Wheeler,
Mertyn Downing, Whitford
Holywell, Flintshire,
CH8 9EP.
01745 560557
mertyndowning@btinternet.
com
GWENYNWYR CYMRAEG
CEREDIGION W.I.Griffiths,
Llain Deg, Comins Coch,
Aberystwyth, SY23 3BG
01970 623334
 wilmair@btinternet.com
LAMPETER AND DISTRICT
Mr Gordon Lumby,
Gwynfryn, Brynteg,
Llanybydder,
SA40 9UX
01570 480571
g.lumby@btopenworld.com

LLEYN AC EIFIONYDD
Amanda Bristow, Bryngwydion,
Pontllyfni, Gwynedd
LL54 5EY 01286 831328
amanda@vosltd.com
MEIRIONNYDD, Lesley Bay,
Hen Orsaf, Gellilydan,
Blaenau-ffestiniog,
LL41 4EP
01766 590488
bazurka@aol.com
MONTGOMERYSHIRE,
Maggie Armstrong,
20 Dol-y-Felin
Abermule
Powys
SY15 6BB
01686 630447
secretary@montybees.org.uk
PEMBROKESHIRE, Brigid Rees,
Canerw Cottage
Llanboidy
Whitland
Carmarthenshire
SA34 0ET
01994 448210
SOUTH CLWYD,
Mrs Carol Keys-Shaw, Y Beudy,
Maesmor Hall, Maerdy
Corwen LL21 0NS
01490 460592
c.keysshaw@btinternet.com
SWANSEA, Paul Lyons,
2 West Cliff, Southgate,
Swansea, SA3 2AN.
paul.lyons@bt.com

176

WBKA/CGC

✉ ☎

TEIFISIDE, Donald Adams,
Bryngwrog
Beulah
Newcastle Emlyn
Ceredigion
SA38 9QR
07932 336076
member@theoldmill.fsnet.
co.uk

WEST GLAMORGAN,
Mr John Beynon,
48, Whitestone Avenue,
Bishopston,
Swansea. SA3 3DA
01792 232810,
jakbeynon@btinternet.com

HEB DADOGU/NON
AFFILIATED:
Mrs J Bromley
Ty Hir, Monmouth Road
Raglan, Usk. NP15 2ET
01291 690331
bromleyjan@hotmail.com

BEIRNIAID SIOE FÊL TRWYDDEDIG / WBKA QUALIFIED HONEY SHOW

TERRY E. ASHLEY
Meadow Cottage,
11 Elton Lane, Winterley
Sandbach CW11 4TN

M. J. BADGER MBE
14 Thorn Lane, Leeds
LS8 1NN

M BESSANT
Gwili Lodge, Heol
Lotwen, Rhydaman
SA18 3RP

ROBERT BREWER
PO Box 369, Hiawassee,
Georgia, USA

TOM CANNING
151 Portadown Road,
Armagh, Co Armagh
BT61 9HL

LES CHIRNSIDE
Bryn-y-Pant Cottage,
Upper Llanover,
Abergavenny NP7 9ES

CARYS EDWARDS
Ty Cerrig, Ganllwyd,
Dolgellau LL40 2TN

IFOR C. EDWARDS
Lleifior, Pontrhydygroes,
Ystrad Meurig SY25 6DN

STEVEN GUEST
Bridge House, Hind
Heath Road, Sandbach,
CW11 3LY

HUGH MCBRIDE
11 Ballyloughan Park
Antrim BT43 5HW

LORRAINE MCBRIDE
11 Ballyloughan Park
Antrim BT43 5HW

CECIL MCMULLAN
33 Glebe Road,
Hillsborough, County
Down

LEO MCGUINESS
89 Dunlade Road, Grey
Steel BT47 4QL

GAIL ORR
64 Ballycrone Road,
Hillsborough BT26 6NH

Graig Fawr Lodge,
Caerphilly, CF83 1NF

REDMOND WILLIAMS
Tincurry, Cahir, Co
Tipperary Eire

MICHAEL YOUNG MBE
Mileaway, Carnreagh,
Hillsborough BT26 6LJ

NBU

NATIONAL BEE UNIT, THE FOOD AND ENVIRONMENT RESEARCH AGENCY

www.nationalbeeunit.com

National Bee Unit
The Food and Environment
Research Agency
Sand Hutton, York, YO41
1LZ, UK

Tel.No: 01904 462510
Fax.No: 01904 462240
E-Mail: nbu@fera.gsi.gov.uk
Website:
www.nationalbeeunit.com
www.fera.defra.gov.uk
Policy : www.defra.gov.uk

NATIONAL BEE UNIT IS NOW UNDER THE FOOD AND ENVIRONMENT RESEARCH AGENCY (FERA)

NATIONAL BEE UNIT

The National Bee Unit (NBU) is part of the executive agency of the Department for Environment, Food and Rural Affairs (Defra), and is based just outside York. The Unit is an element of Fera's Inspectorate Programme and its work covers all aspects of bee health and husbandry in England and Wales, on behalf of Defra in England and for the Welsh Government in Wales. The work of the unit includes disease and pest diagnosis, research into bee health matters, development of contingency plans for emerging threats, import risk analysis, related extension work and consultancy services to both government and industry.

NATIONAL BEE UNIT TECHNICAL
STAFF, HEAD OF UNIT
Mike Brown

HOME BASED STAFF:

NATIONAL BEE INSPECTOR
Andy Wattam
01522 789726
07775 027524

BEE HEALTH INSPECTION SERVICE

The Integrated Bee Health Programme is run by the NBU on behalf of core policy customers. The NBU has a long track record in bee husbandry and bee disease control (since 1946) and has been directly responsible for the bee inspection services in England and Wales since 1994.

The NBU consists of a home-based inspectorate team, and the laboratory diagnostic and research team based at Fera, York. In addition colleagues across Fera contribute to the programme and research projects.

The Bee Health Inspectorate

The inspectorate team consists of approximately 50 home-based members of staff. It is headed by the National Bee Inspector (NBI), whose role it is to manage the statutory disease control and training programmes. The NBI has management responsibility for eight home-

178

based Regional Bee Inspectors (RBIs), one heading each of the seven regions in England and one covering Wales. The RBI in turn manages a number of Seasonal Bee Inspectors (SBIs). The RBIs and SBIs organise inspections under EU and UK legislation, submit suspect samples for diagnosis, treat colonies for foul brood and train beekeepers in bee husbandry for better disease control and greater self-sufficiency. In addition the bee inspectors also collect honey samples for residue analysis under the Statutory Honey collection agreement with Defra Veterinary Medicines Directorate (VMD). With *Aethina tumida* (Small hive beetle (SHB)) and *Tropilaelaps* spp. both notifiable under UK and EU law inspectors also undertake surveillance for these exotics in "at risk apiaries" close to identified high risk areas.

BEE DISEASE DIAGNOSTIC TEAM

The NBU's diagnostic team provides a rapid, modern service for both the inspection team and beekeepers. The NBU laboratory is Good Laboratory Practice (GLP) compliant, a quality accreditation scheme administered by the Department of Health. All diagnostic tests are conducted according to the OIE (Office International des Epizooties) Manual of Standard Diagnostic Tests and Vaccines. The OIE is the world organisation for animal health and produce internationally recognised disease diagnosis guidelines (http//www.oie.int.) Across Fera diagnostic support is provided from teams of microbiologists acarologists, insect virologists and molecular specialists in the Fera Molecular Technology Unit (MTU).

BEES AND THE LAW

The Bees Act 1980 UK empowers Agriculture Ministers to make Orders to control pests and diseases affecting bees, and provides powers of entry for authorised persons. Under the Bees Act, The Bee Diseases and Pests Control Order 2006 for England and Wales, (there is similar legislation for Scotland and Northern Ireland) designates American foulbrood (AFB), European foulbrood (EFB), A. tumida (SHB) and Tropilaelaps mites

REGIONAL BEE INSPECTORS

Ian Molyneux
Northern Region
01204 381186
07775 119442

Charles Millar
Western Region
01694 722419
07775 119476

Nigel Semmence
Southern Region
01264 338694
07776 493649

Alan Byham
South East Region
01306 611016
07775 119447

Adam Vevers
South West Region
01364 653474
07775 119453

Keith Morgan
Eastern Region
01485 520838
07919 004215

Ivor Flatman
North East Region
01924 252795
07775 119436

Frank Gellatly
Wales
01558 650663
07775 119480

FOR DETAILS OF SEASONAL BEE INSPECTORS DETAILS CONTACT THE RELEVANT RBI OR CHECK BEEBASE

179

NBU

LABORATORY BASED STAFF
Research Co-ordinator
Giles Budge

Bee Research
Gay Marris

Laboratory Manager
Ben Jones

Apiary Manager
Damian Cierniak

Administrative
Programme Support
Kate Parker,
Lesley Debenham &
Jenna Cook

(all species) as notifiable pests and defines the action which may be taken in the event of outbreaks.

At the European level, the Directive on animal health requirements for trade in bees is called the Balai Directive (92/65/EEC) implemented in the UK under the Animal and Animal Products (Import and Export) Regulations. It lists American foul brood (AFB), the small hive beetle (A. tumida) and Tropilaelaps mites as notifiable pests and diseases throughout the EU (at the time of writing time neither the small hive beetle nor Tropilaelaps have been confirmed in Europe).

THE IMPORTATION OF BEES

It is legal to import Queen bees from third countries, the rules governing this are set out in Commission Decision 2003/881/EC, as amended by Commission Decision 2005/60/EC. The list of countries is currently restricted to, Argentina, Australia and New Zealand.

It is legal to import bees freely from the EU (including queens, packages and colonies). Under the Balai directive consignments of bees moved between Member States must be accompanied by an original health certificate confirming freedom from notifiable pests and diseases.

For full details on the importation of bees from within the EU or from Third countries please either consult the Defra website, BeeBase or contact the NBU.

AMERICAN AND EUROPEAN FOUL BROOD

Foul brood-infected apiaries are placed under standstill notice, supervised by the bee inspector, until the disease is cleared from the apiary and the honey from antibiotic-treated colonies is safe to harvest. We always aim to minimise the impact of this as far as possible, in co-operation with the beekeeper.

VARROA
As part of the NBU's routine field screening programme the first known case of pyrethroid resistant varroa mites in the UK was discovered in apiaries in Devon in August 2001. The NBU undertook a resistance-monitoring programme throughout England and Wales. Pyrethroid resistant Varroa mites are now widespread in England and Wales. To access current advice on Varroa and Varroa Management please visit BeeBase.

ADULT BEE DISEASES
The NBU also look for adult bee diseases and parasites such as Nosema species (*Nosema apis* and *Nosema. ceranae*, amoeba (*Malpighamoeba mellificae*) and tracheal mites (*Acarine* or *Acarapis woodi*) from samples submitted by beekeepers. As these diseases are non-statutory this service is chargeable. For the current cost please contact the NBU or visit the website.. Bees that have been imported from designated Third countries are also checked for disease and are also screened for exotic pests potentially harmful to UK beekeeping.

EXOTICS
Beekeepers must make themselves aware of the potential threats to beekeeping in the UK. The field inspection team monitors for potential exotics, the SHB and Tropilaelaps spp. The laboratory team also routinely screen import samples and suspect samples submitted for identification by both beekeepers and the field team.

PESTICIDE MONITORING
The Wildlife Incident Investigation Scheme (WIIS) is a unique scheme for monitoring the effects of pesticides on wildlife, including beneficial invertebrates such as honey bees. It is led by the Chemicals Regulation Directorate (CRD) with Natural England managing and undertaking site enquiries on their behalf; The Food and Environment Research Agency (Fera) carry out disease and pesticide analysis and, if appropriate, the Veterinary Laboratories Agency (VLA) carry out post mortems on wildlife. Information gathered is fed into the approval process for pesticides and helps in the verification and improvement of pesticide risk assessments.

It can also result in changes to label recommendations on pesticide products. It is not provided as a personal service to beekeepers wishing to seek evidence for the purpose of civil litigation but can lead to enforcement action being taken by the enforcer if the misuse or abuse of a product is identified as part of this process. For more information please see the website.

RESEARCH & DEVELOPMENT
A programme of research and development within the group underpins the Unit's work. They also have long-established links with many European and world wide research centres, universities and the beekeeping industry. The primary aim of our R&D is to improve our understanding of the issues which impact bee health. The NBU also actively supports PhD students, some of which are funded using donations from the beekeeping industry. For an update on the current R&D work of the unit please see BeeBase.

RISK ASSESSMENT
The National Bee Unit manages 150 honey bee colonies and has much experience in assessing the effects and efficacy of veterinary bee medicines (e.g., varroacides) and pesticides in both field and laboratory tests. Our Good Laboratory Practice (GLP) accreditation allows us to undertake a wide range of routine and specially designed laboratory, semi-field and field studies on honeybees and bumblebees for regulatory authorities and industry worldwide.

EXTENSION
The NBU trains beekeepers in several ways: local courses and advisory visits run by the inspectors, and national courses held at the York laboratory. The NBU annually hosts the National Diploma in Beekeeping residential courses and has also been host to visiting overseas workers and researchers. NBU York based staff also provide training to beekeepers at local and regional beekeeper meetings.

HEALTHY BEES PLAN

The Healthy Bees Plan was published by Defra and the Welsh Assembly Government in March 2009 following consultation with beekeepers and the main Beekeeping Associations. It sets out a plan for Government, beekeepers and other stakeholders to work together to respond effectively to pest and disease threats and to put in place programmes to ensure a sustainable and productive future for beekeeping In England and Wales.
The Healthy Bees Plan consists of three working groups that report to the project management board to help deliver the five major objectives of the plan. To view the Healthy Bees Plan, please see the website.

(This is the most recent information received from the National Bee Unit).

BeeBase is the National Bee Unit website. It is designed for beekeepers and supports Defra, WAG and Scotland's Bee Health Programmes and the Healthy Bees Plan, which set out to protect and sustain our valuable national bee stocks.
Our website provides a wide range of free information for beekeepers, to help keep their honey bees healthy.
We hope both new and experienced beekeepers will find this an extremely useful resource and sign up to BeeBase. Knowing the distribution of beekeepers and their apiaries across the country helps us to effectively monitor and control the spread of serious honey bee pests and diseases, as well as provide up-to-date information in keeping bees healthy and productive. By telling us who you are you'll be playing a very important part in helping to maintain and sustain honey bees for the future.

To register as a beekeeper please visit BeeBase.

DARD

✉ ☎

DEPARTMENT OF AGRICULTURE AND RURAL DEVELOPMENT

www.dardni.gov.uk

BEE DISEASE DIAGNOSTICS:
Sam Clawson
Agri-Food and Biosciences
Institute (AFBI)
Newforge Lane
BELFAST BT9 5PX
Tel: 028 9025 5289
Email: Sam.Clawson@
afbini.gov.uk

TRAINING COURSES:
Jennifer Ball
Greenmount Campus
College of Agriculture Food
and Rural Enterprise:
Information is available
from the College at
Tel: 028 9442 6879
Text phone: 028 9052 4420
Email: Jennifer.Ball@
dardni.gov.uk

BEE INSPECTIONS:
Thomas Williamson
Agri-food Inspection
Branch,
DARD, Glenree House,
Carnbane Industrial Estate,
Newry, Co Down, BT35 6EF
Tel: 028 3889 2374
Fax: 028 3025 3255
Email: Thomas.Williamson@
dardni.gov.uk

Honeybee Regional Report for Northern Ireland 2012

Bee Health Surveys

A questionnaire survey for Bee Husbandry issues has been circulated annually to beekeepers via beekeeping associations since 2009. The results of the 2011 survey are available as a pdf on the AFBI website (www.afbini.gov.uk). This showed colony losses for 2011 were 16% compared to 13% in 2010. Seventy percent of beekeepers reported no losses. The 2012 survey results are currently being processed but will be available on the AFBI website in November.

Bee Health Inspections

The Bee Inspectorate carried out surveys for American foul brood, European foul brood, Small Hive beetle and Tropilaelaps mite along with resistance testing of varroa mites to pyrethroids. American foul brood remains a problem for beekeepers in Northern Ireland with 10 apiaries found to have the disease by early September compared to twelve apiaries in total for 2011. Inspections were also carried out for European foul brood without any incidents recorded. Surveys continued for Small Hive Beetle and Tropilaelaps mite. Apiaries in the vicinity of ports or fruit importers were targeted for Small Hive Beetle inspections using corriboard shelter traps, while apiaries that had imported in the past were selected and hive scrapings examined for Tropilaelaps mite.

184

Varroa

Varroa is ubiquitous in Northern Ireland, consequently no systematic studies of prevalence are conducted. Samples continue to be submitted to monitor varroacide resistance. Six of eleven samples suitable for testing in 2012 returned a positive result for varroacide resistance.

Adult Bee Disease Diagnostics

Nosema ceranae was first recorded in Northern Ireland in 2010. *N. ceranae* is an emergent pathogen of western honeybees. It is similar to the endemic species, *Nosema apis* but is considered to produce a more virulent disease than *N. apis,* probably reflecting its more recent association with the western honeybee. Samples submitted and positive for *Nosema* are screened for *N. ceranae* on an ad-hoc basis. Since April 2011, 142 samples were tested for *N. apis* and *N. ceranae* with 78 (55%) positive for a *Nosema* infection. Of these, 41 (53%) contained *N. ceranae*. Note, however, this should not be used as an indicator of prevalence, as samples were not spatially representative of colony distribution in Northern Ireland.

Up to September 2012, 109 samples of bees have been submitted to the laboratory for disease diagnosis, 33 were positive microscopically for *Nosema* and 15 positive for acarine. In 2011, we had a total of 93 similar samples of which 38 proved positive microscopically for *Nosema* and 13 positive for acarine disease.

Residue Sampling

Honey samples were again lifted this year for testing for residues of veterinary medicines and environmental contaminants. Samples lifted last year were found to be satisfactory.

Imports

Twelve direct Queen imports were notified to DARD in 2012 from Greece and Cyprus. Notification levels have been much less than previous

DARD

years. Follow-up inspections were carried out to check records and health certificates for notified imports.

The Bee Diseases and Pests Control Order (Northern Ireland) 2007
The above Order came into operation on the 21 May 2007, which brought our list of notifiable pests and diseases into line with England and Wales.

Bee Health Contingency Plan
The Bee Health Contingency Plan is reviewed annually and an updated version was published on the DARD Internet in September 2012.

Strategy for the Sustainability of the Honey Bee
The Strategy for the Sustainability of the Honey Bee was published in February 2011 and aims to achieve a sustainable and healthy population of honey bees for both pollination and honey production in the north of Ireland through strengthened partnership working between Government and Stakeholders. The Strategy confirms DARD's ongoing commitment to help protect and improve the health of honey bees and support the sector in its efforts to sustain and support beekeeping. The Ulster Beekeepers Association (UBKA) and the Institute of NI Beekeepers (INIB) have made a commitment to support the Strategy intentions. The Strategy is aimed at both policy makers and beekeepers, and importantly, identifies the roles and responsibilities of the different stakeholders in delivering its aims and outcomes. It seeks to address the current challenges facing beekeepers and provides a plan of action aimed at sustaining the health of honey bees and beekeeping in the north of Ireland for the next decade.

The Strategy for the Sustainability of the Honey Bee can be viewed at:
http://www.dardni.gov.uk/index/publications/pubs-dard-fisheries-farming-and-food/publications-dard-strategy-for-the-sustainability-of-the-honey-bee.htm

Jim Crummie
Head of Crop Certification Plant & Bee Health Inspectorate,
DARD
Thomas Williamson
Senior Crop Certification Plant & Bee Health Inspector, DARD
Sam Clawson
Bee Disease Diagnostics, AFBI
Seamus Hughes
Farm Policy Branch, DARD

SG-AFRC

✉ ☎

THE SCOTTISH GOVERNMENT AGRICULTURE, FOOD AND RURAL COMMUNITIES DIRECTORATE (AFRC) - RURAL PAYMENTS AND INSPECTIONS DIRECTORATE (RPID)

The Scottish Government

HEADQUARTERS
Lead Bee Inspector
Stephen Sunderland,
P Spur, Saughton House,
Broomhouse Drive,
Edinburgh, EH11 3XD
Tel: 0300 244 6672
e-mail: beesmailbox@
scotland.gsi.gov.uk

The Scottish Government (SG) is responsible for bee health Policy in Scotland. SG recognises the importance of a strong Bee health programme, not only for the production of honey, but also for the contribution that bees make to the pollination of many crop species and to the wider environment. Honey bees are susceptible to a variety of threats, including pests and diseases, the likelihood and consequences of which have increased significantly over the last few years.

The Scottish Government takes very seriously any biosecurity threat to the sustainability of the apiculture sector and is working closely with colleagues in Food and Environment Research Agency's (Fera) National Bee Unit (NBU) to enable a more joined up approach to be taken throughout Great Britain on the issues surrounding bee health.

The Scottish Government has invested in the NBU's National web-based database for beekeepers "BeeBase" and actively encourages beekeepers to register onto the system. This service will provide bee health and disease outbreak information and will also assist Bee Inspectors in disease control. BeeBase also provides information on legislation, pests and disease recognition and control, interactive maps, current research areas and key contacts.

Beekeepers have a significant role to play in ensuring disease management and control within their own apiaries are in order as they have a legal obligation to report any suspicion of a notifiable disease or pest to the Bee Inspector at their local SGRPID Area Office. Bee Inspectors are responsible for the operation of The Bee Diseases and Pests Control (Scotland) Order 2007 in their area with duties including:-

- Inspection of apiaries for presence of statutory bee diseases
- Taking and delivering samples to SASA
- Issuing and removal of 'Standstill Notices'
- Issuing of 'Destruction Notices' and supervising destruction
- Informing beekeepers of treatment options for European Foul Brood (EFB), where appropriate
- Granting the option, after taking account of the recommendations of SASA, and carrying out treatment
- Carrying out follow-up inspections after destruction or treatment

SASA

- **Science and Advice for Scottish Agriculture (SASA)** is responsible for providing specialist technical support where duties include:
- Examination of submitted samples suspected of being infected with American Foul Brood, European Foul Brood, Small Hive Beetle (SHB) or *Tropilaelaps*.
- Reporting results on which pathogen or pest is present
- Recommending, in consultation with the Bee Inspector, the most suitable option, destruction or treatment, for each individual case of EFB.
- Where treatment is agreed, ordering supplies of the approved antibiotic
- Provision of a free diagnostic service to beekeepers to identify and confirm the presence of varroa.
- Maintaining technical liaison with NBU and providing technical documentation as required
- Providing training courses and demonstration material as required

SASA (SCIENCE AND ADVICE FOR SCOTTISH AGRICULTURE)
1 Roddinglaw Rd,
Edinburgh, EH12 9FJ

BEE DISEASES,
FIONA HIGHET
Plant Health Section
(0131) 244 8817

PESTICIDE INCIDENTS,
ELIZABETH SHARP
Chemistry Section
(0131) 244 8874

SG-AFRC
✉ ☎

PESTICIDE INCIDENTS

As part of the Wildlife Incident Investigation Scheme (WIIS), SASA undertakes analytical investigations into bee mortalities where pesticide poisoning may have been involved. Beekeepers should send samples of dead bees (200) direct to SASA, Chemistry Section, for analysis. In the case of major incidents, beekeepers are advised to contact their nearest SGRPID Area Office so that an early field investigation can be instigated.

THE FOLLOWING SCOTTISH GOVERNMENT RURAL PAYMENTS AND INSPECTIONS DIRECTORATE (SGRPID) STAFF ARE AUTHORISED BEE INSPECTORS. ALL BEE INSPECTORS HAVE EMAIL ADDRESSES AS "FIRSTNAME.SURNAME@SCOTLAND.GSI.GOV.UK"

EDINBURGH (HQ)
Steve Sunderland
(Lead Bee Inspector)
P Spur, Saughton House,
Broomhouse Drive,
Edinburgh, EH11 3XD
Tel: 0300 244 6672
Fax: 0300 244 9797

GRAMPIAN (INVERURIE AREA OFFICE)
Kirsteen Sutherland
Thainstone Court,
Inverurie, Grampian,
Aberdeenshire, AB51 5YA
Tel: (01467) 626247
Fax: (01467) 626217

SOUTHERN (DUMFRIES AREA OFFICE)
Angus Cameron
161 Brooms Road
Dumfries, DG1 3ES
Tel: (01387) 274400
Fax: (01387) 274440

CENTRAL (PERTH AREA OFFICE)
Kelly Callwood
Strathearn House
Broxden Business Park
Lamberkine Drive
Perth, PH1 1RZ
Tel: 01738 602043

HIGHLAND (INVERNESS AREA OFFICE)
Clem Cuthbert
Longman House
28 Longman Road
Inverness, IV1 1SF
Tel: 01463 253 053

SOUTH EASTERN (GALASHIELS AREA OFFICE)
Angus MacAskill
Cotgreen Road
Tweedbank, Galashiels
Scottish Borders, TD1 3SG
Tel: (01896) 892400
Fax: (01896) 892424

SOUTH WESTERN (AYR AREA OFFICE)
John Smith
Russell House
King Street
Ayr
South Ayrshire
KA8 0BG
Tel: (01292) 291300
Fax: (01292) 291301

GRAEME SHARPE, APICULTURE
Specialist, Veterinary
Services, SAC,
Auchincruive, Ayr.
Tel:01292 525375

SCOTTISH AGRICULTURAL COLLEGE (SAC)

The Scottish Government supports a full-time apiculture specialist (Graeme Sharpe) who provides comprehensive advisory, training and education programmes for beekeepers throughout Scotland on all aspects of Integrated Pest Management, good husbandry (including the control of Varroa) and management practices. SAC also promotes the awareness of notifiable bee diseases and pests.

WWW.SCOTLAND.GOV.UK/TOPICS/APICULTURE/GRANTS/INSPECTIONS/BEEINSPECTIONS

USEFUL TABLES

BEEKEEPING METRIC CONVERTION TABLES

°CENT	FAHR	INCH	MM	INCH	MM	INCH	MM
0	32	$1/25$	1	$1^5/8$	42	10	254
5	40	$1/12$	2	$1^{11}/16$	43	$10^1/4$	260
7	44	$1/8$	3	$1^9/20$	48	$11^1/4$	286
30	86	$1/16$	5	2	51	$11^1/2$	292
34	92	$1/4$	6	3	76	$11^3/4$	298
38	100	$5/16$	8	$4^1/4$	108	12	305
43	110	$3/8$	9	$4^1/2$	114	14	356
49	120	$1/2$	12.5	$4^3/4$	121	$16^1/4$	413
54	130	$5/8$	16	$5^1/2$	140	$16^1/2$	49
60	140	$3/4$	18	$5^3/4$	146	17	431
62	144	$7/8$	22	6	152	$17^5/8$	448
82	180	1	25	$6^1/4$	159	$18^1/8$	460
90	194	$1^1/16$	27	$8^1/4$	216	$18^1/4$	483
100	212	$1^3/8$	35	$8^3/4$	223	20	508
		$1^9/20$	37	$9^1/8$	232	$21^1/2$	546
		$1^1/2$	38	$9^3/8$	239	$21^3/4$	552
				$9^9/16$	246	22	559

INTERNATIONAL QUEEN MARKING COLOURS

YEAR ENDING	COLOUR	REMEMBER
1 & 6	WHITE	Will
2 & 7	YELLOW	You
3 & 8	RED	Raise
4 & 9	GREEN	Good
5 & 0	BLUE	Bees?

USEFUL TABLES

BOTTOM BEE-SPACE HIVES

No, of cells in brood box
Lug length (MM)
Frame spacing (mm)
Frame size (mm)
No. frames
Hive type

Hive type		No. frames	Frame size (mm)	Frame spacing (mm)	Lug length (MM)	No, of cells in brood box
National	BROOD	11	356 x 216	37	38	58000
	SUPER	10	356 x 140	42	38	36000
Modified Commercial	BROOD	11	406 x 254	37	16	75000
	SUPER	10	406 x 152	42	16	

TOP BEE-SPACE HIVES

No, of cells in brood box
Lug length (MM)
Frame spacing (mm)
Frame size (mm)
No. frames
Hive type

Hive type		No. frames	Frame size (mm)	Frame spacing (mm)	Lug length (MM)	No, of cells in brood box
Smith	BROOD	11	356 x 216	37	18	58000
	SUPER	10	356 x 140	42	18	36000
Langstroth	BROOD	10	448 x 232	35	16	68000
	SUPER	10	448 x 140	35	16	
Jumbo	BROOD	10	448 x 286	35	16	85000
	SUPER	10	448 x 140	35	16	
Modified Dadant	BROOD	11	448 x 286	37	16	93000
	SUPER	10	448 x 159	42	16	

USEFUL TABLES

CONVERSION FACTORS

TEMPERATURE

Fahrenheit > Celcius (Centigrade)	- 32, x 0.5555 (⁵/₉)
Celcius > Fahrenheit	x 1.8 (⁹/₅), + 32

WEIGHT

Ounces > Pounds	x 28.3495
Pounds > Grams	x 453.59237
Hundredweights > Kilograms	x 50.8
Grams > Ounces	'/. 28.3495
Kilograms > Pounds	x 2.2142

LENGTH

Inches > Centimetres	x 2.54
Yards > Metres	x 0.9144
Miles > Kilometres	x 1.609
Centimetres > Inches	x 0.3937
Metres > Yards	x 1.0936
Kilometres > Miles	'/. 1.609

AREA

Acres > Hectares	x 0.404686
Hectares > Acres	x 2.47105

VOLUMN

Pints > Litres	x 0.5683
Gallons > Litres	x 4.546
Litres > Pints	x 1.7598
Litres > Gallons	x 0.21997

Notes

www.ingramcontent.com/pod-product-compliance
Lightning Source LLC
Chambersburg PA
CBHW072140270326
41931CB00010B/1832